Print and Tourism

Printing History and Culture

Volume 7

Series Editors

Caroline Archer-Parré, Malcolm Dick and Hazel Wilkinson

PETER LANG

Oxford · Berlin · Bruxelles · Chennai · Lausanne · New York

Print and Tourism

Travel-Related Publications from the Sixteenth to the Twentieth Century

Catherine Armstrong and Elaine Jackson (eds)

PETER LANG

Oxford · Berlin · Bruxelles · Chennai · Lausanne · New York

Bibliographic information published by the Deutsche Nationalbibliothek
The German National Library lists this publication in the German National Bibliography;
detailed bibliographic data is available on the Internet at http://dnb.d-nb.de.

A catalogue record for this book is available from the British Library.

Library of Congress Cataloging-in-Publication Data

Names: Armstrong, Catherine, 1976- editor | Jackson, Elaine, 1955- editor
Title: Print and tourism : travel-related publications from the sixteenth
  to the twentieth century / Catherine Armstrong and Elaine Jackson (eds).
Description: Oxford ; New York : Peter Lang, [2026] | Series: Printing
  history and culture, 2504-4915 ; vol. 7 | Includes bibliographical
  references and index. |
Identifiers: LCCN 2025051461 (print) | LCCN 2025051462 (ebook) | ISBN
  9781803742441 paperback | ISBN 9781803742458 pdf | ISBN 9781803742465 epub
Subjects: LCSH: Travel writing--History | Tourism--History
Classification: LCC G156 .P75 2026 (print) | LCC G156 (ebook)
LC record available at https://lccn.loc.gov/2025051461
LC ebook record available at https://lccn.loc.gov/2025051462

Cover image: Anthony Quinn. *Holidaymaking*, 1956. 56-page magazine-inspired brochure
published by Thomas Cook/Dean & Dawson; printed by LTS Robinson (author's collection).

ISSN 2504-4915
ISBN 978-1-80374-244-1 (Print)
ISBN 978-1-80374-245-8 (ePDF)
ISBN 978-1-80374-246-5 (ePUB)
DOI 10.3726/b20937

© 2026 Peter Lang Group AG, Lausanne, Switzerland
Published by Peter Lang Ltd, Oxford, United Kingdom

info@peterlang.com - www.peterlang.com

Catherine Armstrong and Elaine Jackson have asserted their right under the Copyright,
Designs and Patents Act, 1988, to be identified as Editors of this Work.

This publication has been peer reviewed.

Contact for General Product Safety Regulation (GPSR): gpsr@peterlang.com

Contents

Figures

Tables

CATHERINE ARMSTRONG AND ELAINE JACKSON

Introduction

This volume – part of the 'Printing History and Culture' series – emanated from the Print Networks Conference: *Printing for Tourists*, held in Appleby-in-Westmorland, Cumbria in July 2020. It brings together academic scholars at all career stages and book industry practitioners to explore some key moments in the history of travel writing and publishing through the use of case studies about Britain or British travellers. The chapters are arranged in chronological order, from the sixteenth to the late twentieth centuries. Some of the printed forms discussed, such as Wainwright's *Guides*, will be familiar to scholars, whilst others, such as the nineteenth-century paper peepshows or the Edinburgh fringe fliers, are more unusual forms of touristic print. The emphasis is on the production, distribution and consumption of such items and considers the social and cultural impact that printed ephemera had on particular audiences, dependent on time and place. The volume demonstrates that across a long chronological period, travel and tourism and printing and publishing are intimately bound together.

This book speaks to intersecting scholarly communities working on these four themes: travel and tourism and printing and publishing. The sub-discipline of travel studies, especially in its historical context, has long focused on the way that printed artefacts have resulted from and resulted in travel for a variety of purposes, including exploration, commercial exploitation and leisure. While tourism studies scholars tend to explore the contemporary world and its fascination with travel, book historians have acknowledged the ways that publications about travel in the past have driven the financial and cultural success of many authors, booksellers, printers and later publishers. *Voyages and Visions: Towards a Cultural History of Travel* (1999) by Jaś Elsner and Joan-Pau Rubiés, looked at fiction and non-fiction works alongside film and other cultural artefacts. Jennifer Laing and Warwick Frost use the metaphor, originating with St Augustine, of reading a book being like going on a journey to frame their work *Books and Travel: Inspiration, Quests and Transformation* (2012). The relationship between reader and traveller was not always symbiotic, especially in an imperial context, as illustrated by Mary Louise Pratt in her theorization of the 'travellee' who consumed artefacts of travel writing but did not respond positively to them.[1] Laura Tarkka and Alison Martin develop this idea and show that the line between traveller and travellee, between local, visitor and those remaining at home, was often blurred.[2] Other book trade personnel straddling the divide between reader and author often participated in meaning-making in travel literature, for example translators.[3] In this volume, authors explore the significance of the use of local dialects as a way of fomenting regional pride.

Other authors have more closely focused on the themes of this volume, the relationship between author-traveller and publisher. We argue that publishers and printers acted across professional boundaries too, as did the readers and authors mentioned above, despite the growth of the power of the celebrity author in the nineteenth century. For example, Charles Withers & Innes Keighren examined John Murray's links with the travel writers whose work he published in the early nineteenth century.[4] This volume develops those themes and looks at the networks of British travellers, authors and book trade practitioners across time and place. It examines a range of different print culture artefacts, including ephemera and images,

---

1 Mary Louise Pratt, *Imperial Eyes: Travel Writing and Transculturation* (Routledge: London, 1992).

2 Tarkka, Laura, and Alison E. Martin. 2023. 'Introduction: Travel Writing and the Travellee'. *Studies in Travel Writing* 26 (4): 265–71. <https://doi.org/10.1080/13645 145.2024.2375607>.

3 Martin, A. E. (2014). Outward bound: women translators and scientific travel writing, 1780–1800. *Annals of Science*, 73(2), 157–69. <https://doi.org/10.1080/00033790.201 4.904633>

4 Withers & Keighren, 'Travels into print: authoring, editing and narratives of travel and exploration, c.1815–c.1857' *Transactions of the Institute of British Geographers*, 36/4 (2011), 560–73.

and uses various methodologies, as well as recent digital innovations to access the meaning and significance of those artefacts produced by and for travellers. Through the diversity of its chapters the volume offers a valid contribution to current studies in book, magazine, and print history and gives an insight into social and cultural changes from the sixteenth to the twentieth century. While it concurs with Bill Bell, who argued that the long eighteenth century was a crucial time in the formation of the professional book trades, it traces the evolution of those trades before and since, and the evolution of them in relation to the development of travel, tourism and changing power dynamics across the globe.[5] That evolution, Bell suggests, has never slowed. Despite the decline of British power overseas in the mid-twentieth century, travel writing continued to flourish, until now in the twenty-first century, travel blogs play a significant part in the genre. In addition, students of photography, design, and art history will see how these fields impacted on the promotion of travel and tourism.

Focusing on the sixteenth century, Alex da Costa examines the production of the earliest practical advice given to pilgrims to the Holy Land, including sea travellers, and how printers anticipated and responded to the demand for such texts. The publications discussed include pilgrimage accounts, itineraries, rutters and currency guides as well as those pertaining to any questions the traveller might have on the conduct of the ship's master and crew.

T. M Vozar and Elif Vozar continue with the theme of *ars apodemica* in early modern England taking a digital humanities approach using qualitative data analysis to parse a selection of texts using the software package Nvivo. Their work highlights common themes, formulations of the nature of travel, recommendations, motivations, and contributions to knowledge.

Shijia Yu concentrates on a more visual approach through nineteenth-century English paper peepshows: a paper-based optical toy that was enjoyed by all ages and genders. Yu argues that the paper peepshows

---

5 Bell, Bill. 'The market for travel writing'. *Handbook of British Travel Writing* (2020): 125–41.

depicting landscapes, particularly at the watering resorts of Cheltenham, Brighton and St Leonards-on-Sea can be identified as a 'fancy article' and evidence of conspicuous consumption for tourists.

Continuing the visual theme, Anthony Hamber & Steven F. Joseph examine the rising popularity of topographical and landscape photography in the nineteenth century. Their chapter highlights the versatility of photographs that document, reflect and inform tourist destinations at a time of rapid scientific and societal development.

Ian Maxted focuses on two distinct individuals: Harriette Armytage and John Follett and examines their illustrated travel journals in Devon in the early 1860s. The collections show how cheap steel line-engraved vignettes were used in different forms to illustrate their travels and offer a fascinating insight into the range of images available to tourists at the time.

Susan May examines *The Marvels of Rome* (1889): The First English Translation from Latin of the *Mirabilia urbis Romae*, recognized as an influential medieval description of Rome from the mid-twelfth until the mid-fifteenth century. The chapter is divided into two evoking contemporary attitudes and beliefs. Part one exemplifies some of the archaeological inaccuracies and hagiographical elements in the original manuscripts whereas in part two, the author debates the later and personal influences involved in the choice of the text for translation by Francis Morgan Nichols and its publication by Ellis and Elvey, particularly referencing the late romantics.

Anthony Quinn takes us on 'a whirlwind tour' of British tourism magazines from 1851 to 2020. The chapter identifies the symbiotic relationship between the travel and magazine industries beginning with *The Traveller* (1841) through to when the *Sunday Times Travel* (2020) ceased publication. It covers general magazine articles, the influence of new modes of transport, Victorian titles and niche publications as well as those produced for tourism and travel companies. Included is an examination of changing attitudes to the words 'excursionist', 'tourist' and 'traveller'.

Karen McAulay draws us into the twentieth century to demonstrate how comparatively modest books of Scottish songs: *Scotland Calling*; and

The Glories of Scotland published by Mozart Allan in Glasgow, reflect the publisher's ability to seize a perfect sales opportunity. The publication of the book coincided with a once-in-a-lifetime event, the Festival of Britain in 1951. The author argues that the publisher responded to the rise in tourism to Scotland during the 1950s and that a series of careful decisions were made to 'sell a uniquely Scottish produce to a specific audience in one memorable year', not only in the selection of songs included but also in the accompanying images and its cover presentation.

Where opening chapters in this edition have analysed publications for early pilgrim travellers, Kayla Harris and Sarah Burke Cahalan address the publication of pocket guides to Christian tourism in twentieth-century Britain, post-WWII. Using items from the Marian Library at the University of Dayton, the authors focus on Catholic materials relating to British shrines to the Virgin Mary in Walsingham but also consider practices at Anglican shrines with regard to ecumenism and shared practices that encompassed a broader range of travellers. They reflect that pilgrimages from the United States to Europe at the time were not marketed as strictly religious events but an opportunity to see major landmarks and European cities and demonstrate the interaction between pilgrimages and tourism. The pamphlets discussed include site-specific prayers, listings of restaurants, hotels, souvenir shops and attractions to provide a unique guide for those seeking devotion and tourism.

In the final chapter of this volume, Liz Woodham takes us into perhaps the more familiar territory of Alfred Wainwright's *A Pictorial Guide to the Lakeland Fells*, first published between 1955 and 1966. The author revisits Wainwright's approach to composition and design and argues that despite his description of his guides as 'nothing more than a personal notebook', his work was distinctive because it integrated drawings and texts that directed his readers away simply from 'consuming a printed object' to actually being 'in the landscape with the author'. By analysing the distinguishing elements in Wainwright's composition, design and publication process, the author argues that his work transcends the traditional form of printing for tourists.

1 Practical Advice for Travellers Abroad

While the term 'tourist' would not be coined until the eighteenth century, medieval writers readily understood the pleasures travel had to offer. In a mid-fifteenth-century continuation of *The Canterbury Tales*, a lengthy 'Interlude' describes the pilgrims' arrival in Canterbury and visit to St Thomas' shrine, where they 'pyred fast and poured highe oppon the glasse', passing 'forth boystly, goglyng with hir hedes'.[1] Under the fig leaf of pilgrimage, travellers might act very much like modern day tourists, taking a great deal of pleasure in new sights and experiences. These pleasures might also be had vicariously. While travel abroad was a costly undertaking and often out of reach even for those with means, there was a rich body of works written for would-be pilgrims encompassing itineraries, diaries, letters, devotional aids, guidebooks and travel accounts. By Aryeh Graboïs's count, 135 pilgrimage accounts were written in Europe by the end of the fifteenth century, the majority written in the fourteenth and fifteenth.[2] Continental printers exploited this interest even in the incunable period, at first printing pilgrimage accounts with long manuscript histories, such

1 They looked intently and peered up at the stained glass [...] going forward boisterously, ogling with their heads. Translation my own, based on glosses in 'The Canterbury Interlude and Merchant's Tale of Beryn' in *The Canterbury Tales: Fifteenth-Century Continuations and Additions*, ed. by John M. Bowers (Kalamazoo, Michigan: Medieval Institute Publications, 1992), <https://d.lib.rochester.edu/teams/text/bowers-canterbury-tales-fifteenth-century-interlude-and-marchants-tale-of-beryn>, ll. 149–63, accessed 9 September 2022.

2 Aryeh Graböis, *Le pèlerin occidental en Terre sainte au Moyen Age* (Bruxelles: De Boeck Université Press, 1998), 211–4. Donald Howard offers a larger count of 526 pilgrims writing about travelling to Jerusalem between 1100 and 1500. Donald R. Howard, *Writers and Pilgrims: Medieval Pilgrimage Narratives and their Posterity* (Berkeley, University of California Press, 1980), 17.

as *Mandeville's Travels* and Ludolph von Suchem's *Iter ad Terram Sanctam* but soon progressing to offering works by contemporary travellers. In contrast, it took some time for printers in England to recognize that travellers offered a potential market to be exploited.[3] Nevertheless, in the early sixteenth century, three small works were printed aimed at sea-faring travellers, two for those travelling to Jerusalem and beyond, and a third directed at those going to Compostela in Spain: *Information for Pilgrims unto the Holy Land* (STC 14081, Wynkyn de Worde, 1500), *The Pilgrimage of Sir Richard Guilford*, (STC 12549, Richard Pynson, 1511), *The Pilgrimage of M. Robert Langton Clerk* (STC 15206, Robert Copland, 1522). And by the mid-sixteenth century there was greater recognition of interest among readers in travel with the publication of works like the *Mappa Mundi, Otherwise Called the Compass and Circuit of the World* (STC 17297, *c.*1550), described as 'very necessary for all [...] as wyll labour and traueyle in the countres of the worlde'. In this chapter, I will discuss how printers in England began to recognize, and meet, the demand for practical advice for travellers; how their sense of what the non-professional traveller would desire evolved; and the influence on this on later printed works.

Early Works for Pilgrims Travelling Abroad

Although *Mandeville's Travels* was of no practical use to travellers, being 'desperately out of date' even in the mid-fourteenth century when it first circulated, it was the first pilgrimage account to be printed in England and so a necessary starting point for understanding printed travel literature.[4]

---

3   For a full discussion of the growth of this market in Europe and its influence on English printers see Alexandra da Costa, *Marketing English Books* (Oxford: Oxford University Press, 2020), Ch. 5 'Marketing the Shrine: Pilgrimage Guides, Advertisements and Souvenirs', 165–202.

4   Anthony Bale, '"ut legi": Sir John Mandeville's Audience and Three Late Medieval English Travelers to Italy and Jerusalem', Studies in the Age of Chaucer 38 (2016), 201–37, <https://doi.org/10.1353/sac.2016.0006>. Bale draws attention to a manuscript

Richard Pynson published the first edition in 1496 (*STC* 17246), by which point it was already widely available in print across Europe and he probably expected it to make a reasonable return.[5] In line with English manuscripts of *Mandeville's Travels* – but out of step with the illustrated and well-rubricated continental editions – Pynson's edition was a plain work without woodcuts or chapter titles to divide it and he seems to have relied on the work's existing reputation to sell it.

We cannot know how many copies Pynson sold, but he must have been successful enough to encourage Wynkyn de Worde to think it would be commercially advantageous to print his own, more attractive, edition in 1499 (*STC* 17247). He added rubrics, chapter numbers, a table of contents, and a new incipit that made clear its interest to those wanting to know the way, or imagine travelling to, the Holy Land:

> Here begynneth a lytell treatyse or booke named Johan Maundeuyll knyght born in Englonde in the towne of saynt Albone & speketh of the wayes of the holy londe towarde Jherusalem, & of marueyles of Ynde & of other dyuerse countrees.

While the text was largely the same, the divisions de Worde added drew attention to the end of the prologue and the promise of its usefulness 'specyall for them that wyll & are in purpose for to vysyte the holy cyte of Jerusalem'. The first rubric also invited buyers to think it would furnish them with a helpful itinerary regardless of their means or mode of

---

copy of Mandeville (Manchester, Chetham's Library, MS Mun. A.4. 107, olim MS 6711) once owned by Sir Edmund Wighton which has notes at the back of Wighton's Mediterranean travels. This is 'one of the only pieces of evidence we have to suggest that an English traveller took a copy of *Mandeville* with him on an actual journey' (226).

5 Moseley, 'Whet-stone Leasings of Old Maundevile: Reading the Travels in Early Modern England' in *A Knight's Legacy: Mandeville and Mandevillian Lore in Early Modern England*, edited by Ladan Niayesh (New York: Manchester University Press, 2011), 28–50, 31–2. M. C. Seymour draws attention to the manuscript evidence that Caxton 'once contemplated an English edition' in his 'The Early English Editions of Mandeville's Travels', The Library, s5.19 (1964), 202–7, 202–3, <https://doi.org/10.1093/library/s5-XIX.1.202>.

transport: 'He that wyll go towarde Jerusalem on hors or a foot or by the see' (sig. A2v).

Significantly, de Worde's edition was also accompanied by simplified copies of seventy-two of the woodcuts that Anton Sorg had created for his 1481 Augsburg edition of *Mandeville's Travels*.[6] These woodcuts placed greater emphasis on exotic animals, plants, geological phenomena, and wondrous people (like the Blemmaye), than on religious figures or sacred sites and relics, attempting to capture the reader's interest as much by the thought of the natural marvels of foreign lands as by pilgrimage destinations.[7] These changes clearly gave de Worde the commercial edge for he published two more editions, while Pynson quietly ceded the field.[8]

A year after printing *Mandeville's Travels* for the first time, de Worde tested the market for yet more practical advice for travellers by printing a shorter work called *The Information for Pilgrims unto the Holy Land* in 1500.[9] Just as the incipit for his *Mandeville's Travels* emphasized its interest to those wanting to know 'the wayes' to the Holy Land, this was directed at would-be travellers. Unlike the latter, this short, sixty-page, quarto work was truly pragmatic in its focus and somewhat more up-to-date. It began with the distances between waypoints on the French route to the Holy Land, followed by the German route. After this, the traveller was advised on what coins to take, the exchange rates available along the way and how to make good provision for the sea journey from Venice to Jaffe. This was drawn without acknowledgement from the English portion of William

6 Josephine Waters Bennett, 'The Woodcut Illustrations in the English Editions of *Mandeville's Travels*', *The Papers of the Bibliographical Society of America* 47 (1953): 59–69, <http://www.jstor.org/stable/24299806>. Bennett notes that versions of Sorg's woodcuts are used to illustrate variations of *Mandeville's Travels* in Germany, France, Spain and England until 1722 (69).

7 The Blemmaye were reputed to be headless men: 'theyr eyen are in theyr shulders & theyr mouth is on theyr brest' (fol. 74r).

8 STC 17249, 1503; STC 17248, 1510?

9 Duff, ed., *Information for Pilgrims unto the Holy Land* (London, 1893).

Wey's first *Itinerary*, who made the journey in the late fifteenth century.[10] Next came a translation of an anonymous and rather thin pilgrimage account also undertaken in the late fifteenth century, including lists of the pilgrimage sites at places such as the Vale of Josephat and the Mount of Olives.[11] The account lapsed back into simple Latin for the pilgrimages beyond Jerusalem in Damascus, Mount Sinai and Egypt. Finally, phrases for 'the langage of Moreske & of other countrees also' (sig. e2v), including some Greek and Turkish, and the Stations in Rome, rounded out the slim volume.

De Worde's commercial intuition was correct: there was an appetite for this kind of compilation. The *Information for Pilgrims* went through at least three English editions before 1530 and, as we shall see, some of what de Worde included would become staples of early travel guides: exchange rates, advice on provisioning and useful foreign phrases.[12] The popularity of the *Information for Pilgrims* also seems to have encouraged Richard Pynson to print the *Pilgrimage of Sir Richard Guilford* in 1511. This purported to be the account of Richard Guilford's fatal pilgrimage to the Holy Land in 1506 and was written by the chaplain of Sir Richard's household, Thomas Larke.[13] It was another quarto pamphlet, but at double the length of the *Information for Pilgrims* it offered the reader a more substantial account of the journey and potentially greater insight into what such travel might require: 'This is the begynnynge, and contynuance of the Pylgrymage of Sir Richarde Guylforde Knyght, & controuler vnto our late soueraygne

---

10 George Williams and Bulkeley Bandinel, eds, *The Itineraries of William Wey, Fellow of Eton to Jerusalem, A.D. 1458 and A.D. 1462; and to Saint James of Compostella, A.D. 1456* (London, 1857), 1–7.

11 Duff, ed., *Information for Pilgrims*, ix.

12 Other editions of *Information for Pilgrims*: STC 14082, 1515 and STC 14083, 1524.

13 Rob Lutton, 'Richard Guldeford's Pilgrimage: Piety and Cultural Change in Late Fifteenth- and Early Sixteenth-Century England', *The Journal of the Historical Association* 98 (2013): 41–78, 49 n. 25, <https://doi.org/10.1111/j.1468-229X.2012.00578.x>. See also his 'Pilgrimage and Travel Writing in Early Sixteenth-Century England: The Pilgrim Accounts of Thomas Larke and Robert Langton', *Viator* 48, 333–57.

lorde kynge Henry the.vii. And howe he went with his seruants and company towardes Ihersualem.' While some elements are original, many pages are unacknowledged translations of Bernhard von Breydenbach's *Peregrinatio in terram sanctam* (*ISTC* ib01189000), including the more personal observations.[14]

Only one edition of the *Pilgrimage of Sir Richard Guilford* survives, which may mean it was less popular than the *Information for Pilgrims* despite the apparent attraction of its seemingly more personal elements. On the other hand, as Lutton argues, 'heavy use by pilgrims on the move, as much as lack of popularity might explain the absence of other surviving copies of the *Pilgrimage of Sir Richard Guilford*.'[15] Robert Langton, who published his own itinerary for Compostela (see below) and later travelled to Jerusalem, praised the useful size of the *Pilgrimage of Sir Richard Guilford* – 'a grete light guyde and conducte' – and the level of detail it provided, so that they knew more than the site custodians told them.[16] It may be going too far, as Lutton does, to say that 'by the second decade of the sixteenth century Larke's text had superseded Mandeville's Book as the pilgrim guide of choice for the discriminating English traveller to the Holy Land,' but it was certainly in actual use. Another traveller, Richard Torkington, who left for the Holy Land on 20 March 1517 also appears to have used both the *Information for Pilgrims* and *Pilgrimage of Sir Richard Guilford*, though he does not acknowledge this debt.[17]

---

14   Duff, ed., *Information for Pilgrims*, xi.

15   Lutton, 'Pilgrimage and Travel Writing', 347.

16   Robert Brian Tate and Thorlac Turville-Petre, eds., *Two Pilgrim Itineraries of the Later Middle Ages* (Pontevedra: Xunta de Galicia, 1995), 95. All further page references for Robert Langton's account are to this edition. The prologue refers to 'me and other englysshe pylgrymes that went this yere' to Jerusalem (95) but it is unclear how much time had elapsed between the writing of this preface and its printing.

17   Duff draws attention to Torkington's use of the *Information for Pilgrims* in his edition, xi–xii. See also Julia Boffey, 'Middle English Lives', in *The Cambridge History of Medieval English Literature*, edited by David Wallace (2008), 610–34, 616; and, for a fuller discussion of his borrowings, da Costa, *Marketing English Books*, 182–5.

The problem with the *Pilgrimage of Sir Richard Guilford* was that Guilford's societal position afforded him a rather lavish and privileged experience of the Holy Land, out of reach of many of its readers. Perhaps recognizing this, in 1522, Robert Copland joined Pynson and de Worde in selling pilgrimage guides, but rather than competing directly against their journey guides to the Holy Land, he offered readers *The Pilgrimage of M. Robert Langton, Clerk, to Saint James in Compostella and in other Holy Places of Christendom*. This was a shrewd move by Copland and reflected the extensive interest in journeys to Rome or Santiago de Compostela which were 'shorter, easier and as spiritually beneficial (in terms of indulgences) as the journey to Jerusalem' but without the risks that had arisen since the Ottomans seized Constantinople in 1453.[18] This last, however, began in *media res* at Orleans, passing silently over the short sea journey from England to France. Like *Information for Pilgrims*, the colophon describes it as providing 'for the instruction of good and devout people of Englonde [...] wyllynge to se suche holy places and relykes', such as the names of towns, the leagues and miles between them, and the relics to be found along the way.

Travelling Abroad

Readers of these early pilgrimage accounts would have gained a mixed impression of travel abroad. The greatest attention was given to sea travel, presumably because it was the least familiar and most dangerous means of transport. Nevertheless, contrary to Jean Verdon's assertion that 'for a land civilisation like that of the Middle Ages, the sea could only provoke fear, anxiety, and repulsion,' it was often portrayed as the simple and rather tedious means to an end, with the barest of details offered.[19] Thus, the

---

18   Bale, '"ut legi": Sir John Mandeville's Audience', 206.
19   Jean Verdon, translated by George Holoch, *Travel in the Middle Ages* (Notre Dame, Indiana: University of Notre Dame, 2003), 55.

Information for Pilgrims only mentions that the pilgrims left Venice for Jerusalem at sunset in a merchant's ship and paid according to their estate, while the journey itself is conveyed in the simplest of terms – 'they passed forth', 'they saylled soo forth', 'levynge the londe' (sic. C2v-3r) etcetera – and any detail relates the direction of landmarks, for example: 'they saylled soo forth tyll they came to the yle of Corphu on the ryght honde, & Turky on the lyft honde, eyght myle bytwene both londes' (sig. C2v).

Similarly, there is very little detailed discussion of the river and sea journeys that occupy the first ten folios of *The Pilgrimage of Sir Richard Guilford*. The smooth sea journey to Normandy, for example, is covered by a brief and prosaic note about the place of embarkation, length of journey, and eventual harbour: 'The yere of our Lorde god.M.D.vi. aboute.x. of the Cloke the same nyght we shypped at Rye in Sussex and the nexte daye that was Shyre thursdaye aboute noone we landed at kyryell [Criel-sur-Mer?] in Normandy' (fol. 2r).

For the journey by barge from Tannar to Venice, Larke provides the names of the rivers taken and their forks, and for the galley journey from Venice, he notes the names of ports and sometimes whether the travellers slept on land or on board.[20] There is a palpable sense of the tedium of the outward journey with the effect of the wind receiving the most attention and whether it is 'scarce' or 'easy' (fol. 6v). At Corfona it was 'so scarce and calme' (fol. 7v) that they had to wait for three days before they could land. This tedium is enlivened when the galley stops at various harbours, either religious sites of interest or out of necessity, and Langton remarks about the travellers' disappointment at the isle of Lyssa, where those who did disembark found the village to be 'right lytel worthe' (fol. 7r).

In these accounts, the days at sea are deserving only of more note when things go wrong, such as in the *Information for Pilgrims*, when the ship encounters difficulty near Coron. It is then described as sailing 'vp & downe thre dayes & two nightes in grete peryl besyde grete rockes', with the captain not daring to pass with the wind 'agaynst them'. The sense of

---

20 Galleys were long, light, narrow ships used largely in the Mediterranean. They relied primarily on oarsmen. Verdon, *Travel in the Middle Ages*, 81.

danger is compounded by the mention of one of the rocks being called 'Ouogo' in Greek, or in English 'edgyd' (sig. C3r).

Things go more seriously wrong for Guilford's party on the way home. Having imprudently delayed their return home till October, their ship repeatedly struggles to find a good wind, either unable to make headway or becalmed, so much so that for ten days near Cyprus it cannot sail at all (2–13 October). Worse still, after a further period of stop-start progress, a storm begins on the 24 October at night 'whiche euynynge the wynde began to Inforce and blowe outragyously, and all that nyght Induryd a wondre grete tempest, as well by excedynge over blowynge of wynde as by contynuell lyghtnynge' (fol. 42v). This is the first of many storms that trouble their journey home, resulting in an account that takes up a full seventeen out of sixty folios full of 'pereyll and daunger ... vyolence & Rage' that rather effectively conveys the relentless nature of the voyage to the reader. Whereas the outward journey from Venice to Jaffe took six weeks and three days, they spent nineteen weeks and one day on the return. At several points they fear death, such as when they are unable to find anchor and the ship is driven so near the shore that 'al the maysters and maryners rekenyd nor thought none other but to have ben lost' (sig. H1v). Guilford's party gave thanks to God for the 'grace to have ones the syght of the moost holy lande' (fol. 10r) even after the calm outward journey, but there is a much more palpable sense of relief in their thanksgiving when they reach Venice, delivered many times from 'perysshyng in shypwerke [*sic*]' (fol. 56v).

The understanding that a sea traveller could do nothing in the face of an ill wind without God's grace runs throughout these works. Even with careful preparation safe passage was by no means assured and a comfortable passage even less so. Quite apart from the danger of storms, the discomforts of sea travel listed in the *Information for Pilgrims* include the 'ryght evyll & smouldryng hote and stynkynge' air of the lowest decks, 'rollynge or tomblynge' during a tempest, overcrowding, inferior bread and wine, stagnant water, sea-sickness and diarrhoea (sig. b4r-b5v). Larke describes Guilford's party as enduring many of these on the way home, sometimes having to stay below hatches without relief because of the rain and hail, robbed of sleep by the cries of the sailors and the noise and sight of the storm (fol. 52v).

Nevertheless, these guides do offer advice, both direct and indirect, on how to mitigate some of these evils, with *Information for Pilgrims* providing the most detailed and direct instruction. In it, the reader is told to choose the highest deck if sailing in a galley and to avoid the noxious decks, whereas in a carrack they are advised to choose a space close to the middle of the ship where there is the least rolling during a storm.[21] Pilgrims are advised to be wary of trusting anyone. They are to make sure the owner of the ship swears to fulfil their contractual commitments before the Duke of Venice, on penalty of a thousand ducat fine; to conduct them to harbours along the way where they can get fresh water, bread and meat; to avoid tarrying more than three days at those harbours without the passengers' consent; to not take on further merchandise; to avoid deliberate delay at sea; and to provide hot food twice a day at two meals, with good wine and fresh water (sig. c4v-5r). Nevertheless, the pilgrim is still encouraged to take steps to ensure their own welfare by buying two barrels of wine, each holding ten gallons, and another with water, as well as biscuit, bread, cheese and eggs to tide them over between meals or when food is scant or poor in quality (sig. c5v). To protect these, the pilgrim is instructed to buy a padlock in Venice to secure their chamber when they go on land (sig. b4v) and to store their wine and water within a locked chest because 'yf the shipmen or other pylgrymes may comm therto they wol tame and drynke of it & also stele your water' (sig. b5v).

The Pilgrimage of Sir Richard Guilford does not offer direct advice to future travellers but gives some broad insight into what might be necessary when the pilgrims' preparations for their welfare are mentioned. The reader is told that at Tannar, Guilford's relatives 'stuffed' the barge with vittles, bread and wine (fol. 3r); and that while they followed the coast of Cyprus, the Patron, 'galyottis' (oarsmen?) and pilgrims take in 'wodde, water, beef and moton with all other thynges necessarye' (fol. 10r). The clear pleasure when it is possible to eat 'fresshe vytaylles' on land (fol. 10r)

---

21   The carrack was the largest sail ship in use in the Middle Ages and had much greater carrying capacity than a galley. Castles on either side of the hold provided two floors for passengers to sleep on. Verdon, *Travel in the Middle Ages*, 79.

after even four days at sea is salutary. Larke's work also reminds readers that it may be necessary to make preparations of a different, martial kind. At Corfona the pilgrims are warned of 'Certeyne Turkes Fustis that lay for vs in oure waye', in response to which 'the Patroun of the Galye and every man purveyed to be redy as defensybly as myght be' (fol. 8r).

When on land, securing mounts was an important task. Consequently, among the words the *Information for Pilgrims* provides in Arabic are ass, horse, and mule, while it advised passengers disembarking at Jaffe not to tarry 'for & ye come betyme, ye may chese the best mule or asse that ye can, for ye shall paye no more for the beest than for the worse' (sig. c1r). Perhaps Guilford's party ignored this on the return, since the asses they rode on from Jerusalem to Jaffe were 'right weyke & right symple and evyll trymmyd' (fol. 41r). Larke also found room to complain about the difficulty and 'outragyous Coste' of hiring camels and guides when his master, Guilford, sickened (fol. 12r).

Other Works for Travellers

English printers also began to experiment with printing works for travellers in general inspired by continental examples. For example, short pamphlets with exchange rates and pictures of legal tender helped travellers abroad avoid exploitation and printers were keen to emphasize their wider use beyond mercantile circles. As a Lyon example stated, they were 'grandement profitable pour messeigneurs les marchans et aultres' (1488, *USTC* 76392). The vast majority were printed in Antwerp in Dutch and it must be an edition of *Die evaluacie vanden goude ende selver gedaen int jaer M ccccxcix* that lies behind the only extant English tract of this nature, *The Valuation of Gold and Silver*, which begins with woodcuts of Spanish and Portuguese ducats.[22] It was printed around the time de Worde printed

---

22   According to the USTC, 17 such tracts were printed in Antwerp in Dutch before 1530, four times the number produced by any other city. Peter J. A. Franssen, *Tussen tekst en publiek. Jan van Doesborch, drukker-uitgever en literator te Antwerpen en Utrecht in de eerste helft van de zestiende eeuw.* (Amsterdam, Atlanta GA: Rodopi 1990), 51.

Information for Pilgrims for the first time, which also included briefer information on currency and exchange. *The Valuation of Gold and Silver* claims that it will ensure 'that euery oon of what state or condycion he be shall the better 7 sewerlyer heve the right valure off his golde' (fol. 2r) and would have been of principal use to those travelling in Europe. This kind of practical help had enduring usefulness, even if it needed to be regularly updated. A separate text of a similar kind may have been translated by Laurence Andrew around 1520 and a fragment survives of *The Art and Science of Arithmetic* printed by Richard Fawkes in 1526 which seems to have offered similar content 'to thenten*t* that marchants occupying beyond the see may haue knowledge of there coynes of money that is to say Crones, Ducatz, and Salutz, Fra*n*cz, and with all other small mony after theyr valour'.[23]

A *Routier*, route book or rutter was another kind of work directed at travellers. In 1528, Robert Copland translated and printed for Richard Bankes *The Rutter of the Sea* (*STC* 11550.6), a translation of Pierre Garcie's *Le routier de la mer*, printed at Rouen early in the sixteenth century.[24] This was a navigational aid for those travelling around the coasts of England, Wales, France, Portugal and Spain, describing the 'hauens, rodes, soundynges,

---

   Although the Valuation (STC 24591) is hesitantly attributed to Jan van Doesburch's press and given a date of 1520? by the STC, it was probably the work of another printer in c.1503. The 1499 evaluation would have been rather out of date by 1520 and since the last Dutch edition was printed in 1506 this probably represents the upper limit of its usefulness. Herbert gives the title of another inextant edition that may have been printed later and without reference to the underlying 1499 valuation: *The Valuacion of Golde and Siluer. Made in the famous Citie of Antwarpe, and newely Translated into Englishe by me Laurens Andrewe.* Proctor recognised that William Herbert had conflated the two editions in his *Typographical Antiquities*, though the ESTC does not, attributing the translation of STC 24591 to Laurence Andrew too. See Robert Proctor, *Jan van Doesborgh, Printer at Antwerp: An Essay in Bibliography* (London, 1894), 36–7.

23 London, British Library, Harl. 5919 (178). Not in the STC.

24 D. W. Waters, ed., *The Rutters of the Sea: The Sailing Directions of Pierre Garcie: A Study of the First English and French Printed Sailing Directions* (New Haven: Yale University Press, 1967), 3. See also, *Robert Copland: Poems*, ed. Mary Carpenter Erler (Toronto: University of Toronto Press, 1993), 131–3.

kennynges, wyndes, floodes and ebbes, daungers and costes'.[25] It was the first rutter printed in English and remained the only one in print for fifty years, in contrast to the situation in France where Garcie's short *Le routier de la mer* was swiftly supplanted by his *Le grant routtier*.[26] In his prologue, Copland describes the work as being as essential for the wise 'mayster maryner, or lodesman' as the 'carde, compas, rutter, dyal and other....cables, roopes, ankers, mastes, sayles, ores, artyllery, vytayle, fresshe water, fewell' and other necessities stowed on board (sig. a2v). He also explains that it was brought to him by a 'maryner of the cyte of London' who found a French edition in Bourdieu and thought it ideal for 'al Englyssh men of his faculte [...] [to the] saufgarde of our marchauntes as other hauntynge the se' (sig. a3r). Copland describes the difficulty he had in translating the book because of its obscure navigational and seagoing terms, and how he had to rely on the guidance of experienced mariners who understood its jargon to complete it. It was not then a book he expected the ordinary sea traveller to purchase or fully understand.

Nevertheless, the preface to *Le grant routtier* emphasized the wider-relevance of this work and, by extension, of the shorter *Le routier*, since it contained judgements relating to 'des nefz, des maistres, des mariniers, des marchans and de tout leur estre'.[27] Anyone wishing to travel by ship from England might find value in learning from Copland's translation what would happen to both passengers and cargo if a ship foundered (sig. d6v) or how the master of a ship would respond to drunken and violent behaviour among his crew (sig. d7v). Some of the information it included also had analogues in the works aimed more obviously at non-professional

25 Only the colophon of the first edition survives, all quotations from second edition STC 11550.8 printed by Copland in 1536.

26 For a list of French editions of Garcie's works, see Waters, ed., *The Rutters of the Sea*, 28. Erler lists six English editions of The Rutter: STC 11551, W. Copland, 1555?; STC 11551.5, J. Waley, 1557, STC 11553, T. Colwell, 1560?; STC 11553.3, W. Copland, *c.*1567?; STC 11554 John Awdeley for Antony Kytson, 1573?; and TP 452 in William Ringler, *Bibliography and Index of English Verse Printed 1476–1558* (London and New York: Mansell, 1988).

27 Waters, ed., *The Rutters of the Sea*, 24.

travellers, such as the number of leagues between English and Spanish ports (sig. c6r-v), which may have interested travellers as much as 'The names of the townes and the leyges betwene' in *The Pilgrimage of M. Robert Langton.*

There was then a growing number of works available in English to readers interested in travel abroad by the 1530s, including pilgrimage accounts, itineraries, rutters and currency guides. Unfortunately, the break from Rome of the church in England in 1534, growing evangelical criticism of relics and pilgrimage, as well as the religious and political tensions across Europe in the early decades of the Reformation appear to have made printers wary of continuing to expand on these offerings.[28] More than twenty years elapsed with little travel literature being printed in England. It was not until the 1550s that printers in England seem to have recovered a sense that readers might buy books about travel and the wider world without the excuse of pilgrimage or even of commerce. Two decades after Copland printed *The Rutter of the Sea*, Robert Wyer printed a *Mappa Mundi, Otherwise Called the Compass and Circuit of the World*, described in the colophon as 'very necessary for all Marchauntes and Maryners And for all such, as wyll labour and traueyle in the countres of the worlde'. As the STC notes, and in line with Wyer's general practice of publishing small pamphlets extracted from previously successful works, this material was actually taken from Richard Arnold's *Chronicle* (*STC* 782, 1503?). It described the four quarters of the world, followed by the distances between major cities, covering Europe, parts of the Middle East and North Africa. It thus gave readers a greater understanding of global geography in verbal form.

Of much greater use and interest, however, was Andrew Boorde's *The First Book of the Introduction of Knowledge*, which was first printed in 1542 by William Middleton.[29] Boorde was an itinerant physician who

---

28 For this argument in full see da Costa, *Marketing English Books*, 200–2.
29 The first edition is inextant, but is referred to by Boorde as being in progress – 'a pryntynge besyde saynt Dunstons churche within Temple barre over agaynst the Temple' – in the preface to his *A Compendious Regiment* (STC 3378.5) written the 5 May 1542

spent many years travelling throughout Europe, including to Compostela, and later to Jerusalem and beyond.[30] He wrote extensively about his two great passions, travel and medicine, with his other works including *A Compendious Regiment or a Dietary of Health* (*STC* 3378.5, 1542) and *The Breviary of Health* (*STC* 3374, 1552). Dedicated to Henry VIII's daughter, (then) the Lady Mary, *The Introduction of Knowledge* has thirty-nine chapters, which each deal with a different country, beginning with England, Wales and Scotland. It is described on the title page as

> The fyrst boke of the Introduction of knowledge. The whych dothe teache a man to speake parte of all maner of languages, and to know the usage and fashion of all maner of countreys. And for to know the moste parte of all maner of coynes of money, the whych is currant in euery region.

Most of the places dealt with are European, but the penultimate chapter focuses on Egypt and the last on Jerusalem. Boorde used the conceit of writing for an Englishman travelling from Calais around the circumference of Europe to order the chapters (sig. E2v, M2r), but he seems to have been sincere in his hope that the work would be of help for those who

> doth pretende [i.e. intend] to travayle [...] that they may be in a redines to knowe what they should do whan they come there. And also to know the money of the countre, and to speke parte of the language or speache that there is vsed. (sig. E2r)

Although much of the information that Boorde offered was taken from other sources, there are personal notes that reflect the experiences of his

---

(sig. A2v). The STC identifies this as printed by William Middleton. References here are from the first extant edition, which was printed by William Copland STC 3383, 1555? Copland's second edition is STC 3385, 1562?

30 Elizabeth Lane Furdell, 'Boorde, Andrew (c. 1490–1549), physician and author' in the *Oxford Dictionary of National Biography*, online edition, <https://doi.org/10.1093/ref:odnb/2870>, accessed 9 September 2022.

travels.[31] Some of the most pungent descriptions are of parts of Great Britain. For instance, Boorde has this to say of Cornwall:

> There meate, and theyr bread, and dryncke, is marde and spylt for lacke of good ordring and dressynge [...] there ale is starke nought, lokinge whyte & thycke, as pygges had wrasteled in it, smoky and ropye and never a good sope... (sig. B1v)

In Wales, he recommends avoiding the poor lodging available in the countryside in favour of market towns, where there are 'good vytales, good meate, wine, and competent Ale, and lodgynge' (sig. C1r), while in Scotland, he says there is 'evell ale, excepte Leth ale' and plenty of meat, fish and oat cakes (sig. D1v). But Boorde could comment about places abroad too. He informed the reader, for instance, that in Brabant, in the south Netherlands, there are many 'good fysh and good chepe'; and that in France 'a man shal be honestly orderyd for hys mony and shal have good chere and good lodging' (sig. K1v). His notes gave readers a rough idea of the relative ease or difficulty they might face in different places.

This was not the only book that Boorde wrote with an eye to what would interest potential travellers. As the title of *The Introduction of Knowledge* suggests there was a second volume that is inextant, to which Boorde referred readers who wished to know more of Rome and Italy (sig. H4v). In *The Introduction of Knowledge*, he also stated that he had written a book about 'every region, counte, and provynce' in England (sig. E1v), including the miles, the leagues, and the distance between cities and towns, along with notable facts. Unfortunately, he lent this *Peregrination* to Thomas Cromwell who lost it, though part seems to have survived and was printed in 1735 within a larger compilation.[32] The fact that Boorde

31 Borde describes how he has used *The First Book of the Introduction of Knowledge* to 'shew' the places he has travelled 'specyally aboute Europ, and parte of Afrycke'. He explains that he was never in Asia and so writes about it 'by auctoirs cronycles & by the wordes of credyble parsons the whiche have travelled in those partes' (sig. E1v).
32 In her DNB entry, Furdell draws attention to its survival in T. Hearne, *Benedictus, abbas Petroburgensis, de vita gestis Henrici II et Ricardi I* (1735), 764–804.

wrote three travel guides suggests that he had real confidence in the demand for such information.

It is also striking that Boorde approached travel as a secular endeavour, all his books focus on the kinds of thing that might interest a modern tourist just as much as a sixteenth-century traveller. His *Peregrination* for instance, included lists of sites that would attract a domestic tourist, such as the forests and waters in England, as well as directions for walking round England 'by the townes on the sea coste'.[33] Although key pilgrimage sites, such as Compostela, are mentioned in *The Introduction of Knowledge*, he treats pilgrimage sceptically, as one might expect from someone who worked for Cromwell. For example, he assures the reader that at Compostella 'there is not one heare nor one bone of saint James' and wonders at the clergy who make 'ygnorant people to worship the thyng that is not here' (sig. L4r) and he dismisses Rome as 'an old city [...] greatly decaide' (sig. H4v). Closer to home, he debunks the idea that the waters of St Winifred's Well still run red with her blood:

> There is a wel in wales called saynte wenefrydes well, walshe men sayth that if a man doth cast a cuppe, a staffe or a napkyn in the Well it wyll be full of droppes or frakils and redyshe like bloud, the whiche is false, for I have proved the contrary in sondry tymes. (sig. B4v)

Boorde anticipated that his readers would be interested in the places he discussed for their human rather than spiritual interest and he proved prescient. *The Introduction of Knowledge* was exceptionally popular, with at least one more edition following within five years of the first edition, and a further two by Copland in the next two decades.[34]

---

33 Hearne, *Benedictus*, 800.

34 The STC notes that in addition to the inextant Middleton edition, 'another ed. pr. by R Copland is referred to in 3373.5 and 3386'. In STC 3386, the preface refers to 'a book that is now a pryntyng at Robert cooplondes namyd the Introductyon of knowleg'. I have not been able to consult STC 3373.5 but it was printed in 1547, so within five years at least two (now inextant) editions had been printed, followed by two by William Copland.

Over the course of the first half of the sixteenth century, English print-ers thus moved from printing the intriguing but largely useless *Mandeville's Travels*, to pilgrimage guides with more practical elements, to works directed at the non-professional *and* non-religious traveller, real or imag-inary. Boorde's *The Introduction of Knowledge* represents that step change most fully, though we can also see it in the repackaging of *Mandeville's Travels* by Thomas East as *The Voyage and Travail of Sir John Mandeville* (*STC* 17250, 1568) and by Richard Hakluyt's inclusion of it as one of his *The Principal Navigations, Voyages and Discoveries of the English Nation* (*STC* 12625, 1589). By the mid-sixteenth century, the early experiments of de Worde and Robert Copland had led to a sense of there being an established market for books for travellers and it was possible to talk for the first time of printed books for *tourists*.

ELIF VOZAR AND THOMAS M. VOZAR

2 How to Be a Tourist in the Renaissance: A Qualitative Analysis of Apodemic Themes in English Travel Advice Books, *c.* 1580–1650

The increasing popularity of international travel in the Renaissance, including the development of the more or less standard continental itinerary known as the Grand Tour, was accompanied by a profusion of related printed media from European publishers, including guidebooks and travelogues of all kinds.[1] It also spawned a new genre called the *ars apodemica*, a humanist coinage from Greek *apodēmein* ('to leave one's country'). While other genres of travel writing suggested particular sites abroad or gave vivid descriptions of exotic locales, apodemic texts instead provided general advice relating to travel, ranging from the practical (taking precautions against disease) to the theoretical (why it is worthwhile to observe other nations' customs). While historians of travel have long paid attention to narrative accounts, whether of journeys within Europe or of voyages beyond it, it is only quite recently that travel manuals in the humanist tradition of the *ars apodemica* have begun to be appreciated. Books of this type, which were printed in abundance, have much to offer scholars, not only because they influenced the travel practices of their many readers but also because they offer perspectives on early modern theories of tourism.[2]

---

1 The scholarly literature on early modern travel is too vast to try to encapsulate here, but on the Grand Tour as a Renaissance, rather than a merely eighteenth-century, phenomenon see Edward Chaney and Timothy Wilks, *The Jacobean Grand Tour: Early Stuart Travellers in Europe* (London: Bloomsbury, 2013).

2 On the *ars apodemica* generally see esp. Justin Stagl, *Apodemiken: Eine räsonnierte Bibliographie der reisetheoretischen Literatur des 16., 17. und 18. Jahrhunderts*

This chapter contributes to the growing scholarship on the *ars apodemica* with a study of printed English travel advice from the reign of Elizabeth I to the 1640s. Taking a digital humanities approach, a qualitative analysis of a set of selected sample texts is performed using the software package NVivo, generating insights into the motivations, destinations, recommendations, and risks treated in the apodemic literature of this period. In doing so this chapter shines new light on the intersection of print culture and tourism in early modern England, while also offering a case study for the application of NVivo in humanities research.

NVivo, a qualitative data analysis software package produced by QSR International, is widely used across the social sciences, including such fields as psychology and tourism studies, but remains largely unknown in the humanities.[3] This work therefore constitutes something of an experiment in the application of this social-scientific apparatus to a research topic in the humanities, namely early modern books. In particular, it was determined that this study would employ NVivo to conduct a thematic analysis, defined as 'a method for identifying, analysing and reporting patterns (themes) within [qualitative] data', which it was thought would facilitate

(Paderborn: Ferdinand Schöningh, 1983); Justin Stagl, *Eine Geschichte der Neugier: die Kunst des Reisens 1550–1800*, second edition (Vienna: Böhlau, 2002), published in English as *A History of Curiosity: The Theory of Travel 1550–1800* (London: Routledge, 1995); and Karl A. E. Enenkel and Jan L. de Jong, eds, *Artes Apodemicae and Early Modern Travel Culture, 1550–1700* (Leiden: Brill, 2019).

3 On NVivo as a tool for qualitative analysis see, for example, Patricia Bazeley and Lyn Richards, *The NVivo Qualitative Project Book* (London: Sage, 2000); Nicholas H. Woolf and Christina Silver, *Qualitative Analysis Using NVivo: The Five-Level QDA Method* (New York: Routledge, 2017); and Kristi Jackson and Patricia Bazeley, *Qualitative Data Analysis with NVivo*, third edition (London: Sage, 2019). For an example of the use of NVivo in tourism studies see Elif Vozar, 'Tors, Moors, and Ponies: A Case Study of Stakeholders' Perspectives on Mindfulness and Sustainability at a Nature-Based Tourism Destination in the UK', *Journal of Humanities and Applied Social Sciences* 7/2 (2025), 146–62.

an investigation of various apodemic themes and their distribution in travel advice books.[4]

The first step was the selection of a set of source texts for analysis. The nature of the research entailed several criteria. First, an appropriate source would need to be devoted to the theme of travel. A book like Roger Ascham's humanist education treatise *The Scholemaster* (1570), despite including some discussion of travel to Italy, was therefore excluded. More specifically, a suitable source would need to focus on providing advice for tourists. This eliminated works like William Bourne's *A booke called the treasure for traueilers* (1578), which is concerned with technical matters of navigation, and Joseph Hall's *Quo Vadis?* (1617), which argues against the usefulness of travel entirely. Given that one topic of interest was the distribution of different countries in the sources, books restricted to a single country, such as Robert Dallington's *The View of Fraunce* (1605), were also passed over. Finally, and crucially, the text of the source would need to be markable and machine-readable so that it could be coded and analysed using NVivo. In practice this meant that texts were chosen from the Early English Books Online–Text Creation Partnership (EEBO–TCP) corpus, which features transcriptions from EEBO page images.[5]

This process led to the selection of six texts varying in length from 1,500 to 40,000 words across five books printed between 1592 and 1643: a letter of advice written by Sir Philip Sidney to his brother Robert around 1580, which appeared in the volume *Profitable Instructions* (1633); a letter

---

4 Virginia Braun and Victoria Clarke, 'Using thematic analysis in psychology', *Qualitative Research in Psychology* 3/2 (2006), 77–101, at 79.

5 For discussions about the EEBO corpus, including its origins, its usefulness and its limitations, see Ian Gadd, 'The Use and Misuse of Early English Books Online', *Literature Compass* 6/3 (2009), 680–92; Diana Kichuk, 'Metamorphosis: Remediation in Early English Books Online (EEBO)', *Literary and Linguistic Computing* 22/3 (2007), 291–303; and Bonnie Mak, 'Archaeology of a digitization', *Journal of the Association for Information Science and Technology* 65/8 (2014), 1515–26. See also Michael Gavin, 'How to Think about EEBO', *Textual Cultures* 11/1–2 (2017), pp. 70–105; Peter C. Herman, 'EEBO and Me: An Autobiographical Response to Michael Gavin, "How to Think About EEBO"', *Textual Cultures* 13/1 (2020), 207–16; and Michael Gavin, 'EEBO and Us', *Textual Cultures* 14/1 (2021), 270–8.

of advice to the young Roger Manners, Earl of Rutland dating to circa
1595, also published in *Profitable Instructions*, where it is attributed to
Robert Devereux, Earl of Essex, although a number of scholars now credit
much or all of the letter to Francis Bacon; John Stradling's *A Direction
for Trauailers Taken out of Iustus Lipsius* (1592); Sir Thomas Palmer's *An
Essay of the Meanes hovv to make our Trauailes, into forraine Countries,
the more profitable and honourable* (1606); James Howell's *Instructions for
Forreine Travell* (1642); and Thomas Neale's *A Treatise of Direction* (1643).[6]
Several of these authors are known to have been experienced travellers.
Sidney, for example, had spent several years in France, the Holy Roman
Empire, and Venice, studying at the University of Padua and briefly ven-
turing as far east as Hungary and Poland.[7] Howell, too, made repeated

6   Sir Philip Sidney, 'A Letter to the same purpose', in *Profitable Instructions Describing
    what speciall Obseruations are to be taken by Trauellers in all Nations, States and
    Countries; Pleasant and Profitable* (London: Benjamin Foster, 1633), 74–103. Cf. Roger
    Kuin, ed., *The Correspondence of Sir Philip Sidney, 2 vols.* (Oxford: Oxford University
    Press, 2012), II, 877–82. Robert Devereux, Earl of Essex, 'The Late E. of E. his aduice
    to the E. of R. in his trauels', in *Profitable Instructions*, 27–73. Cf. Alan Stewart with
    Harriet Knight, eds., *The Oxford Francis Bacon I: Early Writings 1584–96* (Oxford:
    Oxford University Press, 2012), 607–50. The true authorship of the letter is irrelevant
    to this study, which is concerned with its circulation in print as a work by Essex. [John
    Stradling], *A Direction for Trauailers. Taken out of Iustus Lipsius* (London: R. B. for
    Cuthbert Burbie, 1592). Sir Thomas Palmer, *An Essay of the Meanes hovv to make our
    Trauailes, into forraine Countries, the more profitable and honourable* (London: H. L.
    for Matthew Lownes, 1606). James Howell, *Instructions for Forreine Travell* (London:
    T. B. for Humphrey Moseley, 1642). Thomas Neale, *A Treatise of Direction, How To
    travell safely, and profitably into Forraigne Countries* (London: Humphrey Robinson,
    1643).
7   See inter alia James M. Osborn, *Young Philip Sidney, 1572–7* (New Haven: Yale
    University Press, 1972); Roger Kuin, 'Philip Sidney's Travels in the Holy Roman
    Empire', *Renaissance Quarterly* 74/3 (2021), 802–28; and T. M. Vozar, 'The Sidney
    Brothers, Elizabethan Travelers, and Hugo Blotius, Prefect of the Imperial Library
    in Vienna', *Sidney Journal* 42/1 (2024), 59–72. On English students at Padua, with
    references to Sidney passim, see Jonathan Woolfson, *Padua and the Tudors: English
    Students in Italy, 1485–1603* (Toronto: Toronto University Press, 1998), with further
    reflections in Jonathan Woolfson, 'Padua and English students revisited', *Renaissance
    Studies* 27/4 (2013), 572–87.

visits to the Continent, including a stay in Madrid during negotiations over the Spanish Match in 1623–4.[8] Palmer, on the other hand, seems to have scarcely stepped beyond his front door: Edward Hasted observed that he 'so constantly resided at Wingham, that he is said to have kept sixty Christmases, without intermission, in this mansion'.[9] Palmer's text was not based on personal experience but was instead freely adapted from Theodor Zwinger's *Methodus apodemica* (1577), just as Stradling's text was, as the title admits, 'taken out of Iustus Lipsius'.

There are many other differences among the sources, including in format – the letters of Sidney and Essex, as previously mentioned, were not standalone books at all but were published together in a single volume along with a synoptic table by William Davison.[10] In a printed context such letters of advice could take on a new life distinct from their epistolary origins.[11] Although the sources have all been grouped together under the genre of the *ars apodemica*, each text has its own emphases and particular concerns. Such discrepancies do not pose a problem for the present research, however, but on the contrary support its goal of attaining a view of the variations in travel advice that print culture made accessible to would-be tourists.

The next stage of research involved uploading the selected texts to NVivo and coding excerpts according to provisional categories that emerged from reading the sources. This was an iterative and to some extent arbitrary process, with categories first developed according to initial impressions, then gradually refined over time. The scope of the

8   See D. R. Woolf, 'Howell, James (1594?–1666)', ODNB.

9   Edward Hasted, *The History and Topographical Survey of the County of Kent: Volume 9* (Canterbury: W. Bristow, 1800), 235.

10 On the use of synoptic tables to organize information about travel, also used by Palmer after Zwinger, see Daniel Carey, 'Inquiries, Heads, and Directions: Orienting Early Modern Travel', in Judy A. Hayden, ed., *Travel Narratives, the New Science, and Literary Discourse, 1569–1750* (London: Routledge, 2012), 25–51, at 32–9.

11 On this point see Elizabeth Williamson, 'A letter of travel advice? Literary rhetoric, scholarly counsel and practical instruction in the *ars apodemica*', *Lives and Letters* 3/1 (2011), 1–22.

analysis could not accommodate consideration of every element in the texts, so it was necessary to concentrate on certain aspects in line with the research aims. Four general themes soon coalesced: 'motivations', or reasons for travel; 'destinations', or places of travel; 'recommendations' or suggested activities; and 'risks', or hazards to avoid. For each theme several sub-themes were sought. While the themes were conceptual in nature, in tracing sub-themes it was found helpful to focus on more objective or practical elements ('meet people', 'learn languages') while largely ignoring the more abstract ('be virtuous', 'control your passions'). Some level of inconsistency, it should be noted, was inevitable: the sub-theme 'Germany' (the Holy Roman Empire), for example, encompasses the Swiss cantons, which were only formally recognized as independent by the Peace of Westphalia in 1648, while the 'Netherlands', despite not being formally ceded by Spain until the same time, constitutes its own sub-theme – but this is how the countries are treated in the sources. A total of eighteen sub-themes were resolved upon, with three or more under each theme. The result was the establishment of a final list of themes and sub-themes, as tabulated (Table 2.1). The coding of the texts was then completed according to this final list, generating the set of textual excerpts for each theme and sub-theme that are outlined in the analysis. A word frequency cloud, produced with a 50-word frequency query based on the coded portions of all six sources (barring less significant lexemes, such as 'and' which were manually marked as stop words), visualizes the most common vocabulary in the data (Figure 2.1).

Analysis

Theme: Motivations

The predominant motivation set forward is the attainment of knowledge, which is highlighted by five of the six texts. Essex acknowledges that some aspects of travel may 'serue for ornaments, al of them for delight', but the

Table 2.1: List of themes and sub-themes

Themes	Sub-themes
Motivations	Knowledge Service Work
Destinations	Italy France Spain Germany Netherlands Elsewhere
Recommendations	Conversation Languages Study Sightseeing Writing
Risks	Outspokenness Vice Health Religion

ultimate motivation should be the cultivation of one's mind: 'the end of Study, Conference, and Obseruation is Knowledge'.[12] Sidney marks the goal of travel as 'the knowledge of such things as may bee serviceable for your Country & calling', 'the right informing your minde with those things which are most notable in those places which you come vnto'.[13] Stradling, similarly, observes: 'Euerie one can gaze, can wander, and can wonder, but to few it is giuen to seek, to search, to learne, and to attaine to true pollicie, and wisedome, (which is traueling indeede.).'[14] Howell declares that the knowledge which comes from direct observation 'is the prime use of Peregrination, which therefore may be not improperly called a moving Academy, or the true Peripatetique Schoole', playing on the literal sense

12 Essex, 'Advice', 31, 65.
13 Sidney, 'Letter', 78–9, 80–1.
14 Stradling, *Direction*, sig. A3v.

Figure 2.1: Total Word Frequency Cloud.

of the Greek *peripatētikos* ('walking about').[15] Neale puts it succinctly: 'The end therefore of discreet Travaile, is Wisedome.'[16] Stradling adduces a plethora of examples of admirable travellers from the Bible, classical antiquity, and English history, but it is worth noting that all five of these texts look to one figure especially as the archetype and ideal of the traveller, namely Odysseus or Ulysses. As Sidney writes: 'who rightly trauels with the eye of *Vlysses*, doth take one of the most excellent ways of worldly wisdome.'[17] In particular, these five authors cite, in Latin or English translation, the same line from the proem of the *Odyssey*: πολλῶν δ' ἀνθρώπων

15   Howell, *Instructions*, 8.
16   Neale, *Treatise*, 1.
17   Sidney, 'Letter', 81.

ἴδεν ἄστεα καὶ νόον ἔγνω ('and many men's cities he saw and mind he came to know', *Od.* 1.3). Odysseus, as one who not only travelled long and far but learned about the peoples whom he encountered, stands a model to be followed by all those who travel seeking knowledge.

Several of the texts also suggest the service of one's country or state as a motivation. Sidney specifies that the knowledge one gains by travel should be 'serviceable for your Country & calling', Stradling that it should 'benefite their owne proper, and natural countrie'.[18] That is also obviously true in the case of government service abroad as a diplomat or soldier as delineated by Palmer, who believes that public benefit should be the motivation of all travel: 'considering of all voluntarie Commendable actions, that of trauailing into forraine States (vndertaken and performed Regularly) is the most behoueable & to be regarded in this Common-weale, both for the publike and priuate good thereof'.[19] Howell at one point declares that the 'most materiall use therefore of Forraine Travel is to find out something that may bee applyable to the publique utility of one's own Countrey' and lists a variety of technologies (masonry, dikes, aqueducts) improved by Englishmen who had travelled abroad.[20] Neale glances at the same idea when discussing potential political entanglements abroad: 'unto this one thing, as to the sole end and Termination, ought all our politicall intentions and actions to be reduced; to wit, that we may profit & benefit our Country.'[21] In one form or another, service constitutes an important motivation across five of the six sources. The more meditative text of Essex is the only exception, though here the service motive may be considered implicit.

Palmer, tellingly, makes no mention of Odysseus. Though he urges the traveller in passing to 'get knowledge for the bettering of himselfe and his Countrie', he emphasizes more prosaic reasons why travel is undertaken, mostly relating to work of various kinds.[22] Some are 'occasioned to

18 Stradling, *Direction*, sig. A3v.
19 Palmer, *Essay*, sig. A1r.
20 Howell, *Instructions*, 208.
21 Neale, *Treatise*, 103.
22 Palmer, *Essay*, 53.

trauell by the Princes or States fauour' as 'men of peace, or men of warre',
the former encompassing diplomats, messengers, couriers, and 'intelli-
gencers' or spies, the latter consisting of soldiers and their commanders.[23]
Merchants will naturally travel for their work as well, as also sometimes
will 'mechanicks' or tradesmen.[24] Clergymen, lawyers, and physicians also
travel for professional reasons. If the other five texts, with their idealization
of Odysseus, are much more humanistically inflected in their focus on the
pursuit of knowledge, Palmer's treatise is by contrast grounded much more
in the practical world of work.

Theme: Destinations

Sidney writes rather dismissively of Italy: 'wee know not what wee haue,
or can haue to doe with them, but to buy their Silkes and Wines.' The
country is full of only 'counterfeit learning'. Politically, 'there is little there
but tyrannous oppression', save Venice, 'whose good Lawes and customes
wee can hardly proportion to our selues, because they are quite of a contrary gouernment'.[25] Stradling, on the other hand, greatly approves of his
addressee's plan to travel to Italy, 'that Queene of countries', where one can
see the places described in Roman historians like Livy and Tacitus and the
physical remains of its ancient empire. Commending especially Venice and
Milan, Stradling adds that Italy can demonstrate good manners ('fine, and
faire cariage of our body, good, & discreet deliuerie of our minde, ciuill,
and modest behauiour to others').[26] Palmer declares that Italy is the most
popular destination of all: 'Italy moueth most of our Trauailers to go and

<hr>

23 Palmer, Essay, 3. On intelligence-gathering as represented in Palmer and other apo-
    demic texts see Elizabeth Williamson, '"Fishing after News" and the *Ars Apodemica*:
    The Intelligencing Role of the Educational Traveller in the Late Sixteenth Century',
    in Joad Raymond and Noah Moxham, eds, News Networks in *Early Modern Europe*
    (Leiden: Brill, 2016), 542–62.
24 Palmer, *Essay*, 19.
25 Sidney, 'Letter', 97–9.
26 Stradling, *Direction*, sig. B4r, C1v.

visit, of anyother State in the world: And not without cause, it being an ancient nurcerie and shop of libertie, the which to the affects of men is precious and estimable.' Among the reasons for this are its climate, its languages, its universities, its manner, its various forms of government ('multiplex and different gouernments, and sundrie policies there found'), and its ancient monuments, though Palmer expresses some scepticism about all of these.[27] Howell, similarly, notes that '*Italy* hath beene alwayes accounted the Nurse of *Policy*, *Learning*, *Musique*, *Architecture*, and *Limning*, with other perfections, which she disperseth to the rest of Europe.' Howell recommends an itinerary passing through Tuscany to Siena, to Milan, to Naples, and above all to Venice, 'a rich magnificent City seated in the very jaws of Neptune', observing that 'there are many things in that Government worth the carying away.'[28] Neale recommends that travellers in Italy visit the universities of '*Rome Bonona*, and *Padua*'.[29]

Sidney names 'France above all other most needfull for vs to marke.' This is especially because of its traditional status as England's enemy ('how they stand towards vs both in power and inclination'), though knowledge of aspects of French government ('the Courts of Parliament, their sub-ulter Iurisdiction, and their continual keeping of payed Souldiers') may also be useful.[30] Stradling supposes that France, like Italy, can teach good manners, and Palmer similarly singles out 'the Courts of France' for offering 'particular gaine of knowledge and information of maners and ciuill cariage'.[31] Howell, like Sidney, sees France as the most sensible choice of destination: 'The first Countrey that is most requisite for the English to know, is France, in regard of neighboured, of conformity in Government in divers things and necessary intelligence of State, and of the use one shall have of that Language wheresoever he passe further.' Howell would have travellers spend some time mastering the language before passing to

---

27 Palmer, *Essay*, 42–3, 44.
28 Howell, *Instructions*, 105–4 [sic; i.e. sig. F4r-v], 108 [sig. F5v], 108 [sig. F6v].
29 Neale, *Treatise*, 32.
30 Sidney, 'Letter', 92, 94–5.
31 Palmer, *Essay*, 42.

Paris, 'that hudge (though durty) Theater of all Nations', where they can study the manners and politics of the French court.[32] Neale recommends that travellers in France visit the universities of 'Paris, and Monpellier'.[33]

Spain, like France, is seen by Sidney as a potential foe of England whose power must be studied ('how they stand towards vs both in power and inclination'), though he also allows that their imperial possessions and other matters ('their good & grave proceedings, their keeping so many Prouinces vnder them, and by what manner; with the true points of honor') may be of interest.[34] Palmer mentions Spain's universities. When in Spain, Howell recommends, the traveller should visit Madrid especially, 'for I know no other place secure enough for a Protestant Gentleman to live in, by reason of the residence of our Ambassador,' but also Seville, the seat of the Casa de Contratación and the end point of the Spanish treasure fleet sailing from the Americas, where one might learn about transatlantic trade.[35] Neale recommends that travellers in Spain visit the universities of '*Toledo* and *Salamanca*'.[36]

Sidney, who had extensive first-hand experience in the country, uniquely sees the Holy Roman Empire above all as a model of good government ('Germany methinks doth excell in good lawes and well administring of Iustice'), adding that travellers should 'consider in it the many Princes with whom we may have league; the places of Trade, and meanes to draw both Souldiers and furniture there in time of need'.[37] Howell mentions in passing the possibility that the traveller returning from Italy may 'crosse the *Alpes*' to Switzerland 'and see some of the *Cantons*, those *rugged Repubiques*, and *Regiments*' as well as 'passe through many of the Stately

32   Howell, *Instructions*, 25, 56.
33   Neale, *Treatise*, 32.
34   Sidney, 'Letter', 94–6.
35   Howell, *Instructions*, 93.
36   Neale, *Treatise*, 33.
37   Sidney, 'Letter', 93–4.

proud Cities of Germany'.[38] Palmer mentions Germany's universities, and Neale specifically recommends the universities of '*Strausburg, and Basill*'.[39]

Sidney mentions 'the Low-Countries' briefly in connection with Spain and notes that Flanders 'hath diuers things to be learn'd, especially their gouerning their Merchants & other trades'.[40] Howell, writing several decades later, regards the Netherlands as a major military power, 'the *very Cockpit of Christendome*', remarkable for its achievements in navigation and commerce, and worthy of visiting for these reasons.[41]

The sources are mostly limited to discussion of travel in western Europe. Howell provides a more far-flung itinerary in the 1650 edition of the *Instructions*, which includes an appendix on the eastern Mediterranean. This appendix could not be formally coded in the analysis, but it is worth discussing briefly here. The focus here is on Constantinople or Istanbul, where the traveller may learn about the Ottoman state ('prying into the errors of [the sultan's] government and weaknesse of his dominions') and the religion of Islam: 'he must observe how it differ's, and in what point in conformes with other Religions' as well as 'how Christians are more beholden to the Turk then to the Jew' in certain beliefs. Greece may be passed through if a land route is taken; otherwise, ships bound for Smyra or Izmir will be found departing Livorno. Beyond the Ottoman capital, Howell suggests that the traveller might visit Cairo and Alexandria in Egypt and stop at Jerusalem on his way back 'such as to see the holy Sepulcher'.[42] Other texts make reference to places outside of Europe, but rarely with the sense that these might serve as possible destinations. Sidney, for instance, notes that it might be useful to learn about the 'good Lawes and Customes' of China, but does not entertain the possibility of travelling somewhere 'which is almost as far as the Antippodes from vs'; Palmer, similarly, makes passing mention of the fortifications of 'Mexico

---

38 Howell, *Instructions*, 161.
39 Neale, *Treatise*, 33.
40 Sidney, 'Letter', 93, 97.
41 Howell, *Instructions*, 162.
42 James Howell, *Instructions*, 130, 132–3, 138.

in West India' and of the culture of countries like Russia and Persia.[43] The exception, to some extent, is Neale's *Treatise*, the epilogue to which states that its advice is addressed especially to 'those, which purpose to travell into Transmarine Regions'. This is the only text of the six that is truly global in scope, with greater attention to places without Europe than within. In a single sentence, to give just one indicative example, Neale can write of 'our young lusty merchants and marriners' travelling in such distant locales as 'Iava, at Bantam; at the Moluccaes, Amboina, Banda, the gulfe of Bengala, Coromandel, Pegu, Tenussery, Mocasser, Achen, Sumatra, Zeilan, and finally in all those hot Countries of China and Iapon'.[44]

Theme: Recommendations

The sources are nearly unanimous in recommending that travellers meet and converse with people worthy of association, particularly those esteemed for their learning, wisdom, and virtue. As Essex puts it, 'your Lordship should rather go an hundred miles to speake with one wise man, than fiue miles to see a fair Towne.'[45] 'But you may say, how shall I get excellent men to take paines to speake with me?', asks Sidney, answering: 'Truly in few words; either much expence or much humblenesse'.[46] Stradling makes the same point by invoking the renowned classical scholar whose *ars apodemica* he has adapted: 'Whoe shall not returne more learned from talking with learned *Lypsius*? a man maye adde to his wisdome verie much, by conferring with the wise saith the wisest of men.'[47] Neale similarly urges travellers 'to get the acquaintance of honest and learned men', enjoining: 'Let therefore a travellour heare of no famous Polititian, or learned

43 Sidney, 'Letter', 90–1; Palmer, *Essay*, 89.
44 Neale, *Treatise*, [166], 12.
45 Essex, 'Advice', 62–3.
46 Sidney, 'Letter', 102–3.
47 Stradling, *Direction*, sig. B3v.

Scholler, but let him endeavour (if he may) to bee his Visitant.'[48] Howell, too, advises conversation with foreigners, but with a focus on learning languages, as when he suggests for the traveller in France 'as a prime help to advance Language, to have some ancient Nunne for a *Divota*, with whom hee may chat at the grates'. Howell also recommends that travellers make it a habit to 'converse with Marchants, and their conversation is much to bee valued, for many of them are very gentile and knowing men in the affaires of the State'.[49] Palmer alone shows reservations about this sort of recommendation when he pointedly disputes the necessity of divines going abroad

> to haue conference with such famous men, whose learning may satisfie & endoc-trine; or else with those naturall Iewes and Grecians whose learning may for the furtherance of those diuine tongues giue much helpe to the vnderstanding of the Scriptures.

This Palmer condemns as a superfluous 'pretence' offensive to England, 'where there are Vniuersities, & publike professors of them'.[50] But these remarks are aimed squarely at absentee clergymen.

Essex remarks only in passing on the opportunity presented to 'learne the language of many Nations'.[51] In the case of Howell, language learning is the overwhelming preoccupation of the entire text, as foregrounded in the title, which promises to show how to 'arrive to the practicall knowl-edge of the languages'.[52] In the main destinations considered (France, Spain, and Italy) the languages are the chief and constant concern, and Howell goes into minute detail about how to build up one's knowledge, how to avoid Anglicisms and other errors, and how different languages

48 Neale, *Treatise*, 42, 159–60.
49 Howell, *Instructions*, 33–4, 98.
50 Palmer, *Essay*, 24–5.
51 Essex, 'Advice', 31.
52 On the learning of Continental vernacular languages in this period England see John Gallagher, *Learning Languages in Early Modern England* (Oxford: Oxford University Press, 2019), with references to Howell passim.

are related to one another. One chapter even offers a long digression on historical linguistics, ending with fulsome praise: 'Speech is the Index, the Interpreter, the Ambassador of the mind, and the Tongue the Vehiculum, the Chariot, which conveyeth and carrieth the notions of the Mind to Reasons Palace, and the impregnable Tower of Truth.'[53] Other sources are less enthusiastic. Sidney admits that 'learning Languages' may 'be of serviceable vse', but stresses that this should not be primary knowledge one seeks, 'for words are but words in what Language soever they be'.[54] Palmer is somewhat ambivalent: in the case of Italian he concludes 'in vaine it is to goe so farre for that, which at home with small paines may singularly bee attained vnto', but elsewhere he urges the traveller to endeavour to perfect his knowledge of languages, which 'is not barely to vnderstand what is read or heard pronounced, but to obserue the peculiar phrase, idiom & construction of words, and singularly to note whereof the tongue hath his speciall deriuation'.[55]

Just as many of the sources agree in recommending conversation with learned men, several also recommend study in foreign universities, schools, and libraries. Stradling advises a stay in '*Bononia* and *Pauia*, the two nurses of Sciences, and liberall arts'.[56] Howell advises one 'to retire to some Vniversity about the *Loire*' in order to learn French properly, or, to the same end, to spend time in 'the *Courts of Pleading*', 'the *Publique Schooles*', or 'the *New Academy*; erected lastly by the *French* Cardinall in *Richelieu*'. Howell notes also the 'divers *Academies* in *Paris*, Colledge-like' where one may 'be taught to Ride, to Fence, to manage Armes, to Dance, Vault, and ply the Mathematiques'.[57] Neale finds the most suitable cities for travellers to visit, 'if thou hast any smacke of learning', to be those 'where there is some *Academy*', listing as examples Paris, Montpellier, Rome, Bologna, Padua, Toledo, Salamanca, Strasbourg, and Basel. Learned associates can

---

53   Howell, *Instructions*, 160.
54   Sidney, 'Letter', 79.
55   Palmer, *Essay*, 43, 54.
56   Stradling, *Direction*, sig. C4v.
57   Howell, *Instructions*, 32–3, 51.

give 'entrance into the most famous libraries, schooles, and Colledges'. 'Let also a travellour', writes Neale, 'passe by no Library of worth, (but if that opportunity may permit) without searching and observing it, committing to memory those things there which he findeth rare'.[58] Palmer is more circumspect. Admitting that travellers find 'residencing in the notable Vniuersities' of Italy appealing, he dismisses these as being 'little beneficiall for a Generalist'. But he allows exceptions for the professions: lawyers might travel 'trauaile into forreine States where there are Vniuersities and where there are degrees to be taken' in order to perfect their vocational knowledge, in which case they should return with the award of a degree as 'the Crowne of their vocation', and physicians might have reason to do the same. In the case of the clergy, however, Palmer understands study abroad as another 'pretence': divines may have in mind 'the view of some famous Librarie, which containeth such famous printed bookes or manuscripts as faithfully discusse of points not yet concluded, nor to bee had and procured other where', but 'transscripts or any assured collections' should render this unnecessary.[59]

Several of the sources promote the value of sightseeing, though others are dismissive. When Odysseus is said to have learned about many cities, Sidney observes, 'hee meanes not (if I be not deceiued) to have seene Townes, and marke their buildings; for surely houses are but houses in every place.'[60] Stradling, by contrast, waxes poetic on the 'so great, and glorious monumentes of antiquitie' to be found in Italy, which 'ingender braue and worthie thoughtes' in the traveller:

> wil he not be greatly delighted with the goodly view of those famous, & delicious places of *Albania, Tibur*, and the renowmed Bathes? What a pleasure will it be to see the house, where *Plinie* dwelt, the countrey wherein the famous *Virgill*, or the renowmed *Ouid* was borne?[61]

<hr>

58 Neale, *Treatise*, 32, 43, 161.
59 Palmer, *Essay*, 43, 26, 24.
60 Sidney, 'Letter', 88–9.
61 Stradling, *Direction*, sig. B4v-C1r.

Palmer recommends attention to 'what manner of buildings are those of Towns & Cities, & of the nobler sort of people' as well as 'the Architecturie of Forts, Townes, Sconces, Cittadels, Castles, Towers, and of places fortified in the land', but nevertheless dismisses 'the speciall gallerie of monuments and olde aged memorials of histories, records of persons and things to bee seene thorowout' Italy as 'a fantasticall attracter, and a glutton-feeder of the appetite, rather than of necessarie knowledge'.[62] Howell, on the other hand, advises that in Italy the traveller 'must note the trace, forme and site of any famous Structure, the Platforms of Gardens, Aqueducts, Grots, Sculptures, and such particularities belonging to accommodation or beauty of dwelling, but specially of Castles, and Fortresses'. Howell later gives a long list of prominent sights throughout Europe:

> To see the Escuriall in Spaine, or the Plate-Fleet at her first arrivall; To see Saint Denis, the late Cardinal-Palace in Richelieu, and other things in France; To see the Citadell of Antwerp; The New Towne of Amsterdam, and the Forrest of Masts, which lye perpetually before her; To see the Imperiall, and stately Hans Towns of Germany; To see the Treasurie of Saint Mark, and Arsenall of Venice'; etc.

Yet all of this, he concludes, like learning languages, 'is but vanity and superficiall Knowledge, unlesse the inward man be bettered hereby'.[63] For Neale, meeting learned men is more important 'then the lofty buildings of the most aspiring Cities', but 'observations of Antiquities, Aedifices, Libraries' can still be rewarding.[64]

Several of the sources mark writing as a necessary part of travel, whether to provide a personal record for one's own recollection in the form of a diary or to present one's experiences to others in the form of letters. Palmer comments that the traveller has a duty to provide information to the English government by 'aduertising, from time to time by Letters during their trauaile, some one of the priuie Councell, and none other of the Countrie to which they belong, of such occurrences and things as chance worthie to be sent and committed to consultation and

---

62 Palmer, *Essay*, 86, 44.
63 Howell, *Instructions*, 109 [sig. F7r], 200–1, 202.
64 Neale, *Treatise*, 160, 154.

viewe'.[65] Howell writes that the traveller 'must alwayes have a Diary about him, when he is in motion of Iourneys, to set down what his Eyes meetes', adding 'for a rule, that Hee offend lesse who writes many toyes, than he, who omits one serious thing'. He further advises that the traveller 'be very punctuall in writing to his Friends once a month at least', insisting that 'there will be plenty of matter to fill his letters withall.' Howell also notes that some keep a 'small leger booke' as an *album amicorum*, 'wherein when they meet with any person of note and eminency, and journey or pension with him any time, they desire him to write his Name, with some short Sentence, which they call *The mot of remembrance*.'[66] Neale notes that it was the practice among the ancients, as it is among 'some few of the most judicious' moderns, to 'commit to their briefe note-books the adventures of each day: and the notable Acts of each weeke to their diaries, Kalendars & Ephemerides'. These may serve as a 'concise magazine' from which the traveller 'might (in *Macrobius* his sence) as from a store-house, drawne out plenty of provision, to put of the famine or barrennesse of oblivion, or their confused memorials'.[67]

Theme: Risks

A prominent concern in several sources is the risk posed by frankness, openness, or outspokenness. Stradling warns of Italian charlatans 'whoe when they haue pried into your nature, & are priuy to your secrets, wil straight change their coppie, and shew themselues in their coulors'. To protect against this he gives as 'three golden rules': '*Frons aperta, Lingua parca, Mens clausa*' (literally 'an open front [or face], a sparing tongue, a closed mind', paraphrased as 'Be friendlie to al, familiar to a few, and speake but sildome').[68] Howell dissuades the traveller from frank speech in Spain especially, writing that he should be 'more reserved and cautelous in his Discours, but entertaine none at all touching Religion, unlesse

65 Palmer, *Essay*, 127.
66 Howell, *Instructions*, 30–1, 53–4, 52–3.
67 Neale, *Treatise*, 26–7.
68 Stradling, *Direction*, sig. C3r-v.

it be with Silence'. He adds that Englishmen, out of 'excessive commendation and magnifyng of his own Countrey', tend 'to undervalue and
vilifie other Countreys, for which I have heard them often censured' –
adding, with a reference to the Spanish Match of then Prince Charles, 'it
had beene wished, some had beene more temperate in this theme at their
being in the Spanish Court, in the yeare 1623'.[69] 'Neither unfitly only doth
the over-nimble discourser in forraigne Countries let slip his words, but
sometimes dangerously', Neale writes, for 'it is imposible for a man which
knoweth not the manners and customes of the men and place before whom
and where he is, to deliver his minde, not to erre and sometimes most
grossely to be deceaved.'[70]

Several sources point to vice in various forms, but particularly sex, as
a danger to the traveller. Stradling warns against 'the alluring and intrap-
ping natures of the Venetian and Italian Curtesanes', asking that he 'take
of me these two precepts: that you refraine your Eyes, and your Eares'.[71]
Howell would not have travellers stay long in Genoa, given among other
reasons the proverbial plenitude there of 'Women without shame'.[72] Neale,
similarly, notes the temptations posed abroad by 'sensuall recreations and
pleasures' of the sort discussed by Ovid and Boccaccio, advising the traveller to actively guard against these.[73]

Several of the sources caution against the risks that travel poses to
one's physical health. Palmer warns that a sudden change of climate, rather
than gradual adjustment, can be 'dangerous' or even deadly. As 'ouer-much
labour distill the *vitall* and *animal* spirits, which is most dangerous', care
should also be taken to ensure proper guidance, clothing, and diet while
travelling.[74] Howell cautions that in Spain one 'must bee much more care-
full of his diet' and 'abstemious from fruit'.[75] Neale, with his global purview,

69 Howell, *Instructions*, 59, 61–2.
70 Neale, *Treatise*, 38.
71 Stradling, *Direction*, C4r.
72 Howell, *Instructions*, 101 [F3r].
73 Neale, *Treatise*, 143.
74 Palmer, *Essay*, 47.
75 Howell, *Instructions*, 59.

is especially concerned with strange or poisonous foods and drugs, such as 'those apples which were found in Guiana by Capt. Vnton Fisher, a little of whose juice causeth sleepe unto death' as well as

> those dangerous Druggs of Petum amongst the Brasilians, Opium amongst the Turkes, Areca and Betelee amongst the Malapars, Cassany rootes amongst the Americans in generall, which are most dangerous to forraigners, and have caused the death of many thousand stout men.[76]

Two sources evince a clear concern that the traveller will be drawn away from true religion, which is understood to lie in the doctrine and discipline of the Protestant Church of England. Palmer writes that the first action a traveller must take upon his return is to 'manifest vnto all men his vncorrupt and vnspotted Religion, and zeale therein'. The maintenance of the orthodox religion is necessary for all travellers, including professionals: 'it is the office of Diuines aboue all things to take heed in their trauaile they be not corrupted with false doctrine', just as it is for lawyers 'to be well grounded in their Religion before, and consequently faithfull'.[77] Howell's advice to the traveller includes 'A caveat touching his Religion', in which he writes that it is 'very requisit that hee who exposeth himselfe to the hazard of Forraine Travell, should bee well grounded and settled in his Religion' so that he may 'returne home an untainted Protestant'.[78]

Conclusion

The relative attention drawn to the respective themes and sub-themes is represented visually in a hierarchy chart (Figure 2.2). In some respects, these results are just as one would expect. Given the popularity of Italian travel, for instance, it is hardly surprising that Italy dominates within the

---

76 Neale, *Direction*, 139–40.
77 Palmer, *Essay*, 129, 25.
78 Howell, *Instructions*, 15–9.

theme 'destinations', with France and Spain not far behind, while Germany, the Netherlands, and other destinations are comparatively insignificant. In terms of motivations, knowledge is preferred overwhelmingly, with service and work reduced to minor possibilities (recreation or pleasure is out of the question, being instead a temptation to the sub-theme 'vice'). This serves to corroborate Karl Enenkel and Jan de Jong's characterization of the authors of early modern apodemic texts as 'intellectuals' who 'reflected upon travelling and tried to discover how it could be optimized as a means of generating knowledge'.[79] If travel, for Renaissance Englishmen, was, at least in theory, chiefly a matter of knowledge and education, then it is fitting to see the extent to which the recommendations in travel advice books are so concerned with conversation, study, the learning of languages, and the reporting of information in writing. That the sub-theme 'conversation' should lead among these accords with the practice of many early modern travellers, including Sidney himself, who made a point of befriending and corresponding with major continental intellectuals like the diplomat Hubert Languet and the botanist Carolus Clusius.[80] Sightseeing makes an appearance, though its appeal is limited.

One finding in particular was quite unanticipated. Under the 'risks' theme, vice, apostasy, and health hazards are more or less predictable threats to a Renaissance Englishman, but the first-ranked sub-theme 'outspokenness' is remarkable in several respects. For one, it is simply not a topic that one would presume to be treated by any travel advice book, Renaissance or otherwise. Moreover, it seems to have received little notice from scholars of early modern travel. Yet this distinctive and coherent anxiety about the dangers of freely speaking one's mind when abroad constituted enough of a concern for three of the six sources to warn against it at length, making

79 Karl Enenkel and Jan L. de Jong, 'Introduction: *Artes Apodemicae* and Early Modern Travel Culture, 1550–1700', in Enenkel and de Jong, *Artes Apodemicae and Early Modern Travel Culture, 1550–1700*, 1–13, at 1.

80 See recently T. M. Vozar, 'An Unpublished Autograph Letter from Sir Philip Sidney to Carolus Clusius, 21 April 1576', *Renaissance Studies* 37/4 (2023), 535–46.

Figure 2.2: Hierarchy Chart.

it the most significant sub-theme in the 'risks' category. There is some sub-stance, it seems, in the advice Polonius gives Laertes as he leaves Denmark for France: 'giue thy thoughts no tongue' (*Hamlet* 1.3.58).[81] It would be worthwhile to explore the ways in which the frankness or outspokenness of the traveller compares to broader early modern notions of *parrhesia* and free speech, though that must be reserved for further research.

As stated at the outset, this study has been somewhat experimental, given the lack of precedents for the use of NVivo in humanities research. This being the case, some reflections on its effectiveness, both for this case and for potential future applications, are due. The present study, co-written by a humanist specializing in early modern culture (T. M. Vozar) and a management researcher familiar with the use of NVivo for qualitative research (Elif Vozar), benefitted from the combination of its

---

81 Cited from Gary Taylor, John Jowett, Terri Bourus, and Gabriel Egan, eds, *The New Oxford Shakespeare: Critical Reference Edition*, 2 vols. (Oxford: Oxford University Press, 2017).

authors' respective expertise in these areas and collaborations of a similar nature are to be encouraged. That said, while NVivo, like any software, requires some degree of orientation before one can adequately employ it, the humanist on the team did not find the learning curve especially steep. Humanities scholars may find NVivo much easier to use than many other digital humanities methods, not least because of its qualitative nature. This is also what potentially makes NVivo a tool well suited to the analysis of historical texts, in parallel with more well-known quantitative methods, and numerous possibilities for research in book history and print culture in particular can be imagined.

Nevertheless, the constraints, while hardly insurmountable, must be taken into account. Early modern orthographical variation, for instance, necessitates some manual manipulation of the word frequency data. Restriction to machine-readable texts limits the objects of study or, alternatively, demands the tedious and time-consuming task of further transcription. The effect of this can be seen in the present study. Specifically, the 'Appendix of som directions for travelling into Turky and the Levant parts' included in the 1650 edition of Howell's *Instructions* could not be coded in NVivo, as only the original 1642 text has been transcribed as part of the EEBO–TCP corpus. The Ottoman Empire, which might otherwise have formed its own sub-theme as a destination, could therefore not be fully integrated into the analysis. This is particularly unfortunate, as the seventeenth century witnessed an increase in travel to the eastern Mediterranean in which Howell could have played a part: in the late 1650s, for example, the young Cambridge scholar Isaac Barrow, later the inaugural Lucasian Professor of Mathematics whose lectures Isaac Newton attended, followed an itinerary from Livorno to Izmir to Istanbul in line with Howell's advice.[82]

---

82   See Howell, *Instructions and Directions*, 130. On travel in the eastern Mediterranean over the long seventeenth century see Gerald M. MacLean, *The Rise of Oriental Travel English Visitors to the Ottoman Empire, 1580–1720* (London: Palgrave Macmillan, 2004). On Barrow's Ottoman travels see T. M. Vozar, 'Isaac Barrow, Ali Ufki and the *Epitome Fidei et Religionis Turcicae*: A Seventeenth-Century Summary of Islam in the European

Despite these constraints, the results of this experimental application should be considered at least moderately successful. The thematic analysis undertaken here has validated in a systematic fashion certain preconceptions about Renaissance travel, including the predominance of Italy as a destination and the prevalence of knowledge as a motivation. Moreover, the analysis has shown how certain specific elements (i.e. the designated sub-themes) fit together across the selection of sample texts and, in the hierarchy chart, has provided a picture of their relative weight. It is hoped that this can form a starting point for future research on the *ars apodemica* and indeed early modern travel more generally. Finally, the analysis has allowed certain previously obscure elements of travel advice books to come into view, in particular the risk posed by frankness or outspokenness. This peculiar concern about free speech has scarcely been noted by scholars of early modern travel, yet the analysis demonstrates that it was highly significant in the source texts.

As with any method, no matter how technical, what one gets out of a software package like NVivo depends on what one brings to it. Nothing is automated, and arbitrary decisions must be taken at every step of the analysis. But under the right circumstances, this study suggests, the use of NVivo for humanities research can serve not only to confirm previous findings but to organize textual data in informative ways and even, potentially, to draw scholarship in new directions.

Republic of Letters', Journal of the *Warburg and Courtauld Institutes* 85 (2022), 145–63 and T. M. Vozar, Isaac Barrow's *On the Turkish Religion: A Latin Poem on Islam from Ottoman Istanbul* (London: Bloomsbury, 2026). Howell is not, however, included in the catalogue of Barrow's library, on which see Mordechai Feingold, 'Isaac Barrow's library', in Mordechai Feingold, ed., *Before Newton: The Life and Times of Isaac Barrow* (Cambridge: Cambridge University Press, 1990), 333–72.

SHIJIA YU

3 Exclusivity in Print: Meaning-Making of Watering Resorts Through Nineteenth-Century English Paper Peepshows

The close relationship between English printed images and tourism has been evolving for centuries, but the period between *c.* 1750 and *c.* 1850 represents a unique phase. On the one hand, this century witnessed a shift in tourists' interest from international to domestic sites, driven by factors such as political turbulence on the Continent in the second half of the eighteenth century and the rising popularity of picturesque tourism, as well as the English urban renaissance and the subsequent process of urbanism.[1] On the other hand, as the art historian Ann Bermingham argues, under the influence of picturesque tours, people's interaction with unfamiliar places was increasingly a visual one, giving rise to the trend of turning travel into a virtual experience realized by consuming pictorial representations.[2]

---

1   Susan Barton, 'Travel Writing, Guides and Journals', in Susan Barton, ed., *Travel and Tourism in Britain, 1700–1914, vol. 1. Travel and Destinations* (London: Pickering & Chatto, 2014), 23. For discussions of the picturesque tourism, see Malcolm Andrews, *The Search for the Picturesque: Landscape, Aesthetics and Tourism, 1760–1800* (Aldershot: Scolar Press, 1989). On the rise of domestic urban tourism, see Katy Layton-Jones, *Beyond the Metropolis: The Changing Image of Urban Britain, 1780– 1880* (Manchester: Manchester University Press, 2016). The English urban renaissance is the term used to describe the changes of the culture and society in English provincial towns between 1660 and 1770. For details, see Peter Borsay, *The English Urban Renaissance: Culture and Society in the Provincial Town, 1660–1770* (Oxford: Clarendon Press, 1991).

2   Ann Bermingham, 'Landscape-O-Rama: The Exhibition Landscape at Somerset House and the Rise of Popular Landscape Entertainments', in David H. Solkin, ed., *Art on the Line: The Royal Academy Exhibitions at Somerset House, 1780–1836* (London: Courtauld Institute of Art, University of London, 2001), 131.

Bermingham refers to such depictions as 'landscape entertainments', and the print market was a major contributor, producing a significant number of images of tourist sites, which catered to a wide range of consumers.[3] Many such print-based landscape entertainments were in novel formats, from the zograscope, which magnifies and enhances the sense of depth in prints, mostly of English urban scenes, to the myriorama that allows for endless combinations of picturesque views, and transparency prints that emphasize the play of light.[4]

The paper peepshow, a paper-based optical toy that started to appear on the English market in 1825, was enjoyed by bourgeois consumers of all genders and ages. Despite receiving little scholarly attention, it is another important print novelty that sometimes assumed the role of a landscape entertainment. Consisting of a front-face with one or more peepholes, several cut-out panels and a backboard lined up one behind another, all connected by bellows on two sides, the paper peepshow presents to its users a miniature three-dimensional immersive world. This format lent itself to representing landscape in innovative ways. Even before the publication of the first English paper peepshow in 1825, *The Areaorama, a View in the Regent's Park*, an amateur maker had already produced what can be considered a proto-paper peepshow, which represents picturesque scenes on the River Wye. Before this optical toy became obsolete in the early 1850s, topography and landscape also continued to constitute its most popular subject matter. The diversity of topics depicted in other paper

---

3 Layton-Jones, *Beyond the Metropolis*, 12.

4 For details on the zograscope, see Erin C. Blake, 'Zograscopes, Perspective Prints, and the Mapping of Polite Space in Mid-Eighteenth-Century England' (PhD diss., Stanford University, 2000). Myriorama consists of a set of illustrated cards that can be arranged in different sequences to form varied views. For details, see Ralph Hyde, 'Myrioramas, Endless Landscapes: The Story of a Craze', *Print Quarterly* 21/4 (December 2004), 403–21. Transparencies were scenes painted on different surfaces, including paper and textile, that were lit on the back. See John Plunkett, 'Light Work: Feminine Leisure and the Making of Transparencies', in Kyriaki Hadjiafxendi and Patricia Zakreski, eds, *Crafting the Woman Professional in the Long Nineteenth Century: Artistry and Industry in Britain* (Burlington: Ashgate, 2013), 44–52.

peepshows, which range from royal ceremonies to the Great Exhibition of 1851, should not be neglected either. Therefore, despite the discussion in this chapter, it would be erroneous to consider the paper peepshow as a product published purely for the purpose of bringing attractive views to users in their homes, as argued by some scholars .[5]

For nineteenth-century consumers, the English paper peepshow portraying landscape and topography operated like other landscape entertainments in the way it functioned to transform sceneries into visual representations for virtual travel. However, it merits in-depth examination due to its various unique features, one of which is the intense focus on watering resorts. Of the thirteen published works that feature landscape or cityscape, seven, all produced between 1828 and 1843, depict inland spas and seaside towns: three show Brighton, two St Leonards-on-Sea, and two Cheltenham, with Cheltenham also featuring in a handmade work. This intriguing phenomenon is the core of my analysis: one crucial function of this collection of paper peepshows of watering resorts was to underpin to their nineteenth-century users the exclusive status of these towns. While the stylized depiction of these places in the cut-out panels would undoubtedly contribute to this construction, existing scholarship has already discussed how visual tropes in the representation of tourist sites invent and reinforce preconceptions of inland spas and seaside towns.[6] Therefore, in my investigation of how the paper peepshow helped frame the perception of watering resorts in the early 1800s, I foreground the role of this print novelty as a commodity and examine the sensations involved in its consumption. Although it is probable that paper peepshows of the towns mentioned here were produced with both local residents and tourists in mind, this chapter examines these works only in relation to the latter

<hr>

5   See for example Francs Terpak, 'Objects and Contexts', in Barbara Maria Stafford, ed., *Devices of Wonder: From the World in a Box to Images on a Screen* (Los Angeles, CA.: Getty Research Institute, c.2001), 341.

6   Peter Borsay's 'A Room with a View: Visualising the Seaside, c.1750–1914', *Transactions of the Royal Historical Society* 23 (2013), 175–201 gives a comprehensive discussion of this topic.

target consumers. First, the role of the paper peepshow as a 'fancy article' is explained by drawing on different aspects of it as a commodity. This feature parallels existing discourse about the prestige of English inland spas and seaside towns in the early decades of the nineteenth century. This chapter then discusses the structure and design of the paper peepshow, and with three works as the primary examples, it uses the theoretical frameworks of remediation and the tourist's gaze to explore what would make these representations of watering resorts particularly effective in constructing exclusivity in print.

Paper Peepshows and Watering Resorts in the Mechanism of Conspicuous Consumption

While archival sources about the nineteenth-century perception of paper peepshows are relatively scarce, they still prove sufficient to conceive of it as a fancy article. To appreciate the significance of this category of commodity, it is necessary to go back to the classical Marxist framework of commodity fetishism, which explains how objects become falsely associated with intrinsic meanings that can then be used to 'stimulat[e] desire that can be realised as market-driven demand'.[7] When the focus of such meanings is placed on identity and status, the idea of conspicuous consumption becomes a natural result. Notwithstanding the fact that this framework was initially developed with the new wealthy classes of the second half of the nineteenth century in mind, the concept proposed by Thorstein Veblen is a helpful framework for the discussion here.

Conspicuous consumption explains the process through which the ostentatious display of certain goods constitutes a crucial part in confirming and making known the social status of their owners.[8] Fancy articles

---

7   Laura Mulvey, *Fetishism and Curiosity* (Bloomington, IN; London: Indiana University Press, 1996), 5.
8   Thorstein Veblen, *The Theory of the Leisure Class* (New York, NY: Dover Publications, 1994).

belong to this group of commodities. While there appears to be no consistent definition for this type of goods in the nineteenth century, references to this phrase can be seen in descriptions of products such as 'work-boxes, turnery, &c., made from a variety of beautiful woods grown in and about the vicinity of Tonbridge'.[9] What these commodities shared is that they were non-utilitarian goods, many of which had been luxury items in the eighteenth century. Often delicately made but sold at an inexpensive price, they might appear trivial, but they were critical to the construction and confirmation of the social identity of their middle-class consumers.[10] In the eighteenth century, such goods, or rather, their predecessors, were the privilege of the genteel class.[11] Often referred to as toys, they were, in fact, sophisticated, even sometimes extravagant ornaments that served to 'decorate the body or distract the mind'.[12] For the economist Adam Smith, the upper classes' pursuit of these goods was the childish behaviour from which middle-class economic rationality should be distinguished.[13] In the chapter on toys in *Practical Education*, co-written with her father Richard Lovell Edgeworth, Maria Edgeworth also stresses the importance of teaching children to distinguish the so-called good toys – useful educational tools – from bad toys, which are frivolous and useless goods, namely fancy articles.[14] Yet these cautionary notes did not prevent the middle classes from developing an interest in such goods, as they gained the connotation of upper-class identity and taste through the process of commodity fetishism. Consequently, in the early nineteenth century, such non-utilitarian

---

9   John Bruce, *The History of Brighton with the Latest Improvements*, 4th ed. (Brighton: John Bruce; London: John van Voorst, 1835), x.

10 Tammy Whitlock, *Crime, Gender and Consumer Culture in Nineteenth-Century England* (Aldershot: Ashgate, 2005), 52.

11 Whitlock, *Crime, Gender and Consumer Culture in Nineteenth-Century England*, 26.

12 Melinda Alliker Rabb, *Miniature and the English Imagination: Literature, Cognition, and Small-Scale Culture, 1650–1765* (Cambridge: Cambridge University Press, 2021), 28.

13 Teresa Michals, 'Experimenting Before Breakfast: Toy Education and Middle-Class Childhood', in Dennis Denisoff, ed., *The Nineteenth-Century Child and Consumer Culture* (Aldershot; Burlington, VT: Ashgate, 2008), 34.

14 Michals, 'Experimenting Before Breakfast: Toy Education and Middle-Class Childhood', 29–42.

novelties were part of the goods purchased in acts of conspicuous consumption since they were vital to the middle classes' claim to status and capital, as they filled their parlour and announced their owners' financial capability and awareness of fashion.[15]

The paper peepshow fits the criteria of a fancy article perfectly. Usually sold for a price between five and seven shillings, it was out of reach for the working classes but was by no means a luxury item for the middle classes, who would happily give out much more for other print-based landscape entertainments.[16] At the same time, its delicate paper materiality and miniature form hint at its lack of practical purposes apart from providing amusement. Moreover, a handful of advertisements from publishers, which are all that we have in terms of nineteenth-century representations of the paper peepshow, consistently frame their product as a pastime designed to provide entertainment. Unlike many other public and domestic visual recreations of the same period, no reference is ever made to any educational value of the paper peepshow. This categorization can be further evidenced by the fact that many vendors of this print novelty were known for their close association with fancy articles. A label that repeatedly appears on English paper peepshows is that of a certain W. & A. Essex, the occupier of stands number 333, 4, 5 and 6 in the Soho Bazaar.[17] Opened by John Trotter in 1816, the Soho Bazaar was the first of its kind in London.[18] Apart from being a novel shopping institution for the well-off, known for the exotic qualities derived from its name, bazaars were also sometimes noted or critiqued for being places where fancy articles were sold.[19] More specific to the discussion in this chapter, the vendor of one work portraying

15 Whitlock, *Crime, Gender and Consumer Culture*, 26; 52.
16 For instance, as noted in Hyde, 'Myrioramas, Endless Landscapes: The Story of a Craze', a myriorama could cost fifteen shillings.
17 For one example of such a work, see Ralph Hyde, *Paper Peepshows: The Jaqueline & Jonathan Gestetner Collection* (Woodbridge: Antique Collectors' Club, 2015), 186.
18 Jane Rendall, '"Bazaar Beauties" or "Pleasure is our Pursuit": A Spatial Story of Exchange', in Iain Borden, ed., *The Unknown City: Contesting Architecture and Social Space* (Cambridge, MA; London: MIT Press, 2002), 112.
19 Whitlock, *Crime, Gender and Consumer Culture*, 47.

Brighton, Daniel Harding Greenin, maintained a shop selling fancy articles and toys such as Tunbridge ware and dolls' house furniture.[20] Henry Lamb, the publisher responsible for two works on Cheltenham, also ran a fancy repository that sold various materials for art-making alongside 'a great variety of fancy articles'.[21]

As the commodity that marks the status of its owner, or in other words, as a key player in the mechanism of bourgeois conspicuous consumption, the paper peepshow has much in common with English watering resorts in the early decades of the 1800s. While the practice of using mineral water for its medical properties had been in place in England for hundreds of years, spa-visiting as a privileged activity for the royals, aristocrats and landed gentries only gained significant popularity in the eighteenth century.[22] The health benefits constituted just one part of the motive for going to inland resorts, as the element of leisure and pleasure was also an essential part of spa-visiting. They functioned not only as places where upper-class visitors could recuperate their physical well-being but also as alternative locations to London for their social season, a new type of town for recreation.[23] By the mid-nineteenth century, inland spas had been transformed from tourist attractions into residential towns.[24] The rise of English seaside resorts came slightly later, in the second half of the eighteenth century. Although the origin of sea bathing lies in the traditional practices of inhabitants near the sea, it took a long time for an appreciation of the sea to be established.[25] Long feared for its wild waves and potential to engulf human civilization, the seascape only started to be considered approachable from the eighteenth century onwards, and

20 See Hyde, *Paper Peepshows*, 197, for more information on Greenin's business.
21 Steven Blake, *Views of Cheltenham 1786–1860: Topographical Prints of a Regency Town* (Cheltenham: Cheltenham Art Gallery & Museums, 1984), 25.
22 Phyllis Hembry, *The English Spa, 1560–1815: A Social History* (London: Athlone Press; Rutherford, NJ: Fairleigh Dickinson University Press, 1990), 1–3; 111.
23 J. A. R. Pimlott, *The Englishman's Holiday* (London: Faber and Faber, 1947), 23–7.
24 John K. Walton, *The English Seaside Resort: A Social History, 1750–1914* (Leicester: Leicester University Press, 1983), 6–7.
25 Walton, *The English Seaside*, 10.

positive associations with the sea, which included the health benefits of sea bathing, also began to form during this period.[26] Starting from the 1750s, such new perceptions of the coast contributed to the transformation of many seaside towns into a new type of fashionable resort, first for the upper classes and then for the middle classes, modelled after inland spas.[27] From the early nineteenth century, coastal towns gradually replaced inland spas as the preferred watering resorts and they sustained their popularity for a longer period, having adapted themselves to the age of mass tourism in the second half of the nineteenth century.[28]

For the upper and middle classes, the provision of recreation at watering resorts was at least as important as that of health recuperation. The venues that specialized in enabling activities of high-status leisure, including assembly rooms, circulating libraries, walks and squares, were common in inland spas and seaside towns.[29] Major resorts would also often have a Master of Ceremonies, who would ensure that those participating in these activities obeyed the codes set out by the privileged society.[30] Considered in this context, visiting watering resorts could thus function as a non-material expression and confirmation of one's status and capital. In other words, it was a form of conspicuous consumption, as only those from the more affluent part of society would know the appropriate manners and behaviours for these towns.[31] In addition, the leisure activities commonly practised in watering resorts also offered opportunities for the material manifestation

26 See Alain Corbin, *The Lure of the Sea: The Discovery of the Seaside in the Western World, 1750–1840*, Jocelyn Phelps, trans., (Berkeley; Los Angeles, CA: University of California Press, 1994), 1–53 for an extensive discussion of this in the context of Western Europe.

27 Walton, *The English Seaside*, 12. See page 16 in the same volume for a discussion of the major differences between these two types of resorts.

28 The transformation of English seaside resorts after the 1850s is beyond the scope of the discussion of this chapter. For detailed analysis, see for example Walton, *The English Seaside*, 156; 162.

29 Walton, *The English Seaside*, 12; 112–3; Peter Borsay, 'A Room with a View', 179–80.

30 Walton, *The English Seaside*, 12; 112–3.

31 Borsay, 'A Room with a View', 179–80.

of conspicuous consumption by encouraging the display of luxury goods that symbolized a high social standing.[32]

Between the 1820s and 1840s, when all the paper peepshows discussed here were produced, Cheltenham, Brighton and St Leonards-on-Sea were among the most popular and fashionable watering resorts, so it is unsurprising that they appear to be the preference of publishers. Located in Gloucestershire, Cheltenham's first mineral spring, which became the Old Well, was discovered in the early eighteenth century. The Old Well Walk, a tree-lined avenue leading to the spring, which appears in both works examined here, was laid out in 1739.[33] The Walk was one of Cheltenham's first walks and rides intended for a fashionable promenade. However, it was only at the end of the century that Cheltenham began to develop into a spa resort of national importance, thanks to the royal patronage of George III of the Old Well in 1788.[34] Its accelerated growth started at the turn of the century and culminated in the 1820s and 1830s.[35]

The history of Brighton follows a similar path, although, in the early decades of the nineteenth century, it occupied a much more prominent position among watering resorts compared to the other two towns. Dr Richard Russell, a resident in Brighton in the mid-eighteenth century, played an instrumental role in the development of this town. By advocating the medical benefits of sea bathing with his treatise *A Dissertation Concerning the Use of Sea Water in Diseases of the Glands*, he effectively endorsed the sea bathing attraction of Brighton and contributed to its evolution from a fishing village to watering resort.[36] With the patronage of the Prince Regent, later George IV, its development was phenomenal. The Brighton Royal Chain Pier, the focus of all paper peepshows of Brighton, was part of it. Constructed in response to the need to offer the Dieppe packet boats a landing place, the potential of the Chain Pier as a venue for

32 Borsay, 'A Room with a View', 179.
33 Hembry, *The English Spa*, 179–81.
34 Hembry, *The English Spa*, 191–2.
35 Hembry, *The English Spa*, 199–201.
36 Martin Easdown, *Piers of Sussex* (Stroud: History, 2009), 63.

a pleasure promenade was discovered.[37] It soon overtook the position of the Old Steine as the central location for social interaction and conspicuous display and became one of the icons of Brighton in the 1820s.[38]

St Leonards-on-Sea in East Sussex on the southeast coast of England came onto the scene of watering resorts relatively late. In the late 1820s, the architect James Burton, who played a crucial role in many Regency-style architectural and urban design projects, most famously the Regent's Park, decided to build his own coastal town near Hastings. While Burton had intended the new town primarily as a prestigious location for the upper classes who would stay for the long term, it differed little from other resorts and was also popular with middle-class visitors.[39] Although only finished in 1828, St Leonards-on-Sea quickly attracted much attention and was regarded by some nineteenth-century observers as potentially more desirable than Brighton.[40]

The parallel between the paper peepshow and watering resorts in enabling the middle classes to display and reinforce their status is thus clear. In paper peepshows of inland spas or seaside towns, these two forms of expression of one's social standing converged into one. Possessing one such item could simultaneously demonstrate the owners' appreciation for the significance of resort visiting and their awareness of the latest fancy articles. At the same time, the repeated appearance of watering resorts in the paper peepshow effectively reinforced the prestige of the resorts by confirming them as the fitting subject matter for this new fancy article. In fact, the connection between fancy articles and inland spas and seaside towns appears to be already established in some of these resorts in the early nineteenth century. Apart from the two examples of Greenin and Lamb

37 Easdown, *Piers of Sussex*, 63.

38 John Ford and Jill Ford, *Images of Brighton* (Richmond-upon-Thames: Saint Helena Press, 1981), 29; 41.

39 J. Manwaring Baines, *Burton's St. Leonards* (Hastings: Hasting Museum, 1956), 9–11. See also Elizabeth Nathaniels, 'James and Decimus Burton's Regency New Town, 1827–37', *The Georgian Group Journal* 4 (2012), 162, where she argues that St Leonards-on-Sea was the example of 'a bourgeois society with aristocratic yearnings'.

40 Anonymous, *The Watering Places of Great Britain and Fashionable Directory* (London: Joseph Robins, 1833), 104–5.

mentioned above, in the guidebook for Cheltenham, for example, it is written that in the New Market House on the High Street of the town, there was 'a neat and very elegant range of shops, in which fancy goods are principally sold'.[41] A similar situation can be observed in Brighton, which was allegedly a town whose business 'consists chiefly in the manufacture of fancy articles'.[42] The selling of such goods was also common here, as many directories and guidebooks record how the interiors of the towers of the Royal Chain Pier were 'neatly fitted up for the sale of books, prints, confectionery and other fancy articles'.[43] It is possible that producers of paper peepshows were aware of this association between watering resorts and fancy articles and used it purposefully when choosing the subject matter of their works.

Reinforcing Prestige in the Peep-View

The paper peepshow is not only effective in highlighting the exclusive status of inland spas and seaside towns as a fancy article; its structure and design reinforce this. On a superficial level, compared to many other print formats, the paper peepshow is special as it allows only one user at a time to peer into and be immersed in the world behind the front-face. In this sense, the experience of using such a print novelty to appreciate watering resorts aligned well with the discourse of exclusivity. Also, the emotions of tourists nearing these towns could be activated by the experience of using the paper peepshow, thereby underscoring the prestige of these places. The analysis concentrates on the works *Cheltenhamorama, a View of the Old Well Walk* and *Interior View of Brighton Royal Chain Pier* (see Figures 3.1 and 3.2).[44]

---

41 Samuel Young Griffith, *Griffith's New Historical Description of Cheltenham and Its Vicinity*, vol. 1, 2nd ed. (Cheltenham; London: Longman, Rus, Orme, Brown & Green, 1826), 15.
42 Bruce, *The History of Brighton with the Latest Improvements*, x.
43 Anonymous, The Watering Places of Great Britain, 22.
44 *The Cheltenhamorama, A View of the Old Well Walk*, Henry Lamb, *c.*1832, Gestetner 226; *The Cheltenhamorama, A View of the Old Well Walk*, Henry Lamb, *c.*1832,

Figure 3.1: Cat 227 *The Cheltenhamorama, a View of the Old Well Walk*. Published by Henry Lamb. Hand-coloured lithograph. 15 x 10.8 x 69 cm (expanded). *c.*1832. Peep-view. Gestetner 227, Jacqueline and Jonathan Gestetner Collection, Victoria and Albert Museum, London. © Courtesy of the Victoria and Albert Museum, London.

The similarities in design between paper peepshows of watering resorts mean that it is possible to make general arguments based on a select group of works. Moreover, the works chosen here illuminate the comparison between the paper peepshow and other media, because it is likely that their publishers derived the peep-views from topographical prints of conventional formats.

The case with *Cheltenhamorama, a View of the Old Well Walk* is straightforward. The publisher Henry Lamb was an artist who also lithographed prints himself and he opened his first shop, a fancy repository, at the latest by 1824 at 98 High Street, Cheltenham.[45] Around the same time, he published the first of his two sets of prints, *Views of Cheltenham and its Vicinity*, consisting of at least seventeen views.[46] As his business continued to grow, by 1825, Lamb had already opened a second repository and moved his main premises to 421 High Street.[47] He produced the second set of prints of the same title in 1833, with fewer works.[48] In both sets, an image of the Old Well can be found, which has a very similar composition to the peep-view of the two *Cheltenhamorama, a View of the Old Well Walk*.

While recycling materials across different formats of printed matters was not a new practice by the early 1800s, there are many factors that could have contributed to Lamb turning his attention to paper peepshows. As one of the most popular inland spas of the 1820s and 1830s, Cheltenham received a significant number of fashionable visitors from London, who perhaps brought information about the latest developments in visual culture, including the appearance of the paper peepshow. In addition, the first set of Lamb's prints was produced in London, another way that this new

---

Gestetner 227; *Interior View of Brighton Royal Chain Pier*, Anonymous, *c.*1829, Gestetner 215. All works are at the Victoria and Albert Museum, London. As the two *Cheltenhamorama* works look very similar, in the discussion of the content of the two paper peepshows, I will refer to them as one work, unless otherwise stated.

45 Blake, *Views of Cheltenham 1786–1860*, 5; 20.

46 Blake, *Views of Cheltenham 1786–1860*, 20.

47 Blake, *Views of Cheltenham 1786–1860*, 25.

48 Blake, *Views of Cheltenham 1786–1860*, 24–5.

Figure 3.2. Cat 215 *Interior View of Brighton Royal Chain Pier*. Anonymous.
Hand-coloured aquatint. 11 x 14 x 54 cm (expanded). *c*.1829. Peep-view. Gestetner 215,
Jacqueline and Jonathan Gestetner Collection, Victoria and Albert Museum,
London. © Courtesy of the Victoria and Albert Museum, London.

print novelty travelled from the capital to the province.[49] Access to the
paper peepshow would thus have been easy for Lamb. In addition, since
he had fancy articles in stock at his repository, Lamb was probably aware
of the commercial potential of such goods. The combination of Lamb's
expertise, his connection with London, as well as the scope of his business,
would have thus provided the perfect environment that inspired him to
convert an existent print into paper peepshows.

Establishing the connection between *Interior View of Brighton Royal
Chain Pier*, a work whose publisher is unknown, and a particular topographical print is more complicated. While the work bears no publisher's

49 Blake, *Views of Cheltenham 1786–1860*, 20.

imprint, the peep-view is strikingly similar to a print by John Bruce by the same title, as part of his *Select Views of Brighton*, first produced in 1827, with subsequent editions in 1828, 1829 and 1833. While this period witnessed the production of numerous visual representations of the Pier, many of which have the same composition as the one in the paper peepshow, Bruce's is most similar.[50] In Fig 3.2, two groups of figures in the paper peepshow are clearly derived from Bruce's work. On the first cut-out panel, on the left in the foreground, a woman wearing a pink hat and dress holding the arm of a man with a top hat and brown coat encounters a man in military uniform. This group of three can also be seen in the image by Bruce, although the arrangement of figures is now in reverse. On the same panel on the right, the man in a top hat and blue coat with a walking stick is almost certainly the same figure who appears in Bruce's print in approximately the same position, although the person speaking to this man is now different. In addition, the paper peepshow and the work from Bruce have the same title, which is not used on other topographical images of the Chain Pier.[51]

Considering the similarities in the composition, arrangement of figures and title, it is likely that, like Lamb, Bruce himself reworked his print into a paper peepshow. Some indirect evidence can further support this hypothesis. Although Bruce did not run a fancy repository and sold fancy articles like Lamb, he would know this category of goods and their commercial potential too. It was in Bruce's guidebook on Brighton that the particularly prosperous business of fancy article manufacture in the town was mentioned, and he also displayed his prints for sale in the Chain Pier's tower, where many shops selling fancy articles were allocated.[52] Brighton was also influenced by the London print world. For example, Rudolf Ackermann, the vendor of the paper peepshow *The Areaorama, a View*

---

50 See Ford and Ford, *Images of Brighton*, Gallery 294–316, for a survey of the prints depicting the Pier Head, which is the angle taken by the cut-out panels in *Interior View of Brighton Royal Chain Pier*.

51 See Ford and Ford, *Images of Brighton*, 289–91, for titles of the prints depicting the Pier Head.

52 See Ford and Ford, *Images of Brighton*, 71.

in the Regent's Park, set up a local branch in Brighton in the 1830s. It is thus understandable how information about this optical toy travelled to this town too.[53] In addition, Bruce was a versatile artist-publisher and he advertised himself in the directories as an engraver, an artist and finally a publisher.[54] He had also produced a few illustrated guides to Brighton and was thus familiar with topographical images in formats other than the conventional print.[55] Knowledgeable about the production process of topographical prints and other formats of scenes of Brighton, Bruce was in a position to access information about novel trends in the print market while also having the propensity to engage with prints of Brighton.

While the refashioning of topographical prints into paper peepshows does not indicate a hierarchy between these two media, at least one person appeared to be convinced that the latter would sell better, despite the already crowded market of topography prints in both towns, which was full of products with similar motifs of well-known sites.[56] Lamb sold the first of his two sets of *Views of Cheltenham and its Vicinity*, which contained at least seventeen views, at *7s 6d*, exactly the price indicated on one version of *The Cheltenhamorama, a View of the Old Well Walk*, which only provided one peep-view.[57] Why would customers be willing to pay the same amount of money for fewer views? One important reason could be the ability of paper peepshows to bring their users closer to watering resorts more effectively by evoking the sensation middle-class tourists would likely experience when approaching these towns, a feeling that would also reinforce the exclusive nature of these places. While this is not restricted to works whose content recycles conventional prints, the

53   Andy Grant and Steve Myall, *Victorian Chroniclers of Brighton* (Brighton: Regency Society of Brighton and Hove, 2017), 79.
54   Ford and Ford, *Images of Brighton*, 135.
55   Apart from the guide referred to note 9, Bruce has also published Bruce's History of Brighton and Stranger's Guide (Brighton: John Bruce, 1828).
56   See Blake, *Views of Cheltenham 1786–1860* and Ford and Ford, *Images of Brighton* for details on the variety of prints produced in the early 1800s on Cheltenham and Brighton.
57   Blake, *Views of Cheltenham 1786–1860*, 20.

comparison between paper peepshows and conventional prints is clearer when the visual design has much in common.

My examination of the ability of the two media to trigger emotions is not based on arbitrary criteria but draws on the framework of remediation and, more specifically, the concepts of immediacy and hypermediacy. In their conceptualization of remediation, Jay David Bolter and Richard Grusin argue that immediacy and hypermediacy constitute the double sides of this process within and between media.[58] They theorize that in any given medium, the logic of immediacy strives to erase the medium's presence so that we feel as if we are actually in the presence of the represented object, while hypermediacy makes visible and even highlights the process of mediation.[59] Despite their different agendas, immediacy and hypermediacy are both manifestations of the desire to achieve the real, which is the goal of remediation. Here, the real is not understood in the metaphysical sense, but defined in terms of what the viewer/user experiences as authentic.[60] In the context of immediacy, this means that our experience of the represented object is considered real since the fact of mediation is erased; while for hypermediacy, the real is achieved as we acknowledge the mediation process and take our experience of the medium itself as something authentic.[61]

Bolter and Grusin argue that while the double logic co-exists in all media, immediacy nonetheless occupies the dominant role, as can be demonstrated by the development of western visual representation since at least the Renaissance.[62] Users of different media also desire a higher level of immediacy, a wish most clearly manifested in the process of a new medium remediating an old one. Bolter and Grusin contend that the new needs to justify its existence by proving its ability to fulfil an unkept

---

58   Bolter and Grusin, *Remediation: Understanding New Media* (Cambridge, MA.: MIT Press, 2001), 5.
59   Bolter and Grusin, *Remediation*, 5–6.
60   Bolter and Grusin, *Remediation*, 53.
61   Bolter and Grusin, *Remediation*, 53; 70–1.
62   Bolter and Grusin, *Remediation*, 24; 34.

promise of the old, which is often about a lack of immediacy.[63] While they also observe that in some media, such as nineteenth-century optical devices including the kaleidoscope and the stereoscope, hypermediacy instead of immediacy is primarily responsible for the appeal of the media, they regard this phenomenon as an exception.[64] They argue the fact that these devices ultimately lost their popularity and even became forgotten is evidence that the desire for hypermediacy is only temporary, whereas achieving immediacy is what people always wish for.[65]

According to the logic of remediation, as a new visual representation of watering resorts compared to conventional topography prints, the paper peepshow showcased a higher degree of immediacy, and this feature helped underpin the significance of the watering resorts. This chapter expands the definition of Bolter and Grusin. In their original argument, the erasure or reduction of traces of mediation on the represented object constitutes immediacy, as this allows the viewer to be brought closer to or directly in front of what is depicted. But this argument has a too strong focus on the visual and does not pay enough attention to the other senses or emotions, which contribute to essential aspects of our experience of objects or events.[66] When media reproduce, simulate or evoke elements such as the sound or feelings associated with the represented entity, the distance between it and the viewer can also be reduced, which is another form in which immediacy can be achieved.

This type of immediacy concerning emotions and sentiments is the focus of this analysis. Neither *Cheltenhamorama, a View of the Old Well Walk* and *Telescopic View of the Chain Pier, Brighton*, nor the prints these works derived from, could effectively erase the traces of mediation for their nineteenth-century users. In the prints by Lamb and Bruce, linear

63 Bolter and Grusin, *Remediation*, 60.
64 Bolter and Grusin, *Remediation*, 37–8.
65 Bolter and Grusin, *Remediation*, 37–8.
66 This is also the case when the two scholars discuss examples of immediacy. However, in the analysis of hypermediacy, the non-visual aspects are given more attention. See Bolter and Gruisin, *Remediation*, 71–2.

perspective is the main technique used to reduce the impression of representation. The trees alongside the Old Well Walk and the towers of the Pier form orthogonal lines that are effective in creating the impression of depth and space to generate the sense that the surface of the image is dissolved and that viewers could project themselves into the depicted scene. The figures portrayed in both prints, with their back facing us, functioned as surrogate figures for the viewers, who could be more effectively taken into the picture. In paper peepshows, the three-dimensional structure is responsible for realizing perspective in a material way. As users let their eyes go into the depth of the cut-out panels and wander among the scenes represented on them, they could experience an immersive sensation, which would more effectively enable them to imagine themselves in the landscape portrayed than topographical prints could. Yet the resulting advantage over the experience afforded by topographical prints was limited. The immersive sensation would not function as the reiteration of the travelled experience for users, nor render the traces of mediation invisible. The images of the cut-out panels, like the prints, are executed in a gestural way without many fine details, which is a style that does not conform to the visual convention of realism of the nineteenth century and shows clear marks of representation. Moreover, when a paper peepshow is expanded, the cut-out panels often have gaps between them when seen through the peephole. Users thus needed to activate their imagination to help complete the peep-view so that a coherent scene could form in their eyes, resulting in an even less realistic impression of the peep-view.

Therefore, in comparison to topography prints, paper peepshows of watering resorts did not offer a more naturalistic depiction, nor could they bring users significantly closer to the scenes by denying traces of mediation. However, paper peepshows would be able to prove to be more effective in evoking the emotions associated with the experience of approaching inland spas and seaside towns as tourists, thereby achieving a higher level of immediacy in this way. The conceptualization by the sociologist of tourism and mobility, John Urry, which has been further extended by other scholars, provides an important theoretical framework for explaining why the sensations evoked would be of significance and how they would be related to the prestige of watering resorts. Urry argues that the core of tourism is about experiencing the extraordinary.[67] Building on this idea, he proposes the concept of the 'tourist gaze'. This notion refers to a way of looking, focused on seeking experiences not encountered in everyday life, which form a mechanism essential to the way tourism operates. Not only does the 'tourist gaze' direct the eyes to unfamiliar features of landscape and townscape, but it also provides a set of discourses that shapes how visitors perceive and interpret what they see.[68] Following this framework, it becomes clear that by repeatedly depicting watering resorts according to a particular aesthetic paradigm, visual representations such as topographical prints and paper peepshows both play a critical part in the formation of the 'tourist gaze', as they enable it 'to be endlessly reproduced and recaptured'.[69] The large number of prints that utilize similar compositions of landmarks in watering resorts manifests such a gaze.

While Urry's formulation concentrates on how the nature of tourism affects the way sites are perceived and interpreted, the historian Peter Borsay expands Urry's theory and argues that the journey to the destination should also be considered as influenced by the 'tourist gaze'.[70] Through an examination of eighteenth- and nineteenth-century travel literature, Borsay highlights how the association of tourism with the unfamiliar that demands a different way of looking was manifested when tourists were close to their destination. Describing their excitement of seeing the end of their journey, tourists often constructed a strong contrast between the scenes they saw earlier on the trip and the sight in front of them, highlighting the impression that the latter was a completely different world while

---

67 John Urry, *The Tourist Gaze 3.0* (Los Angeles, CA: SAGE, 2011), 1. For some of the subsequent literature discussing the idea of tourism as an activity of being away from the everyday, see for example Hartmut Berghoff and Barbara Korte, 'Britain and the Making of Modern Tourism: An Interdisciplinary Approach', in Hartmut Berghoff et.al., eds, *The Making of Modern Tourism: The Cultural History of the British Experience, 1600–2000* (Basingstoke: Palgrave, 2001), 3.

68 Urry, *The Tourist Gaze 3.0*, 3.

69 Urry, *The Tourist Gaze*, 3.

70 Borsay, 'A Room with a View', 175–201.

also expressing their heightened anticipation.[71] In part, tourists might develop this feeling due to a sense of disorientation. Since the beginning of the nineteenth century, travel times had been significantly reduced by extensive building of turnpike roads and the improvement of the coach.[72] The swift relocation of tourists could generate bewilderment, as they went into the coach with their experience of their home city and encountered a different world when they stepped out some hours later.[73] Yet, at the same time, the build-up of the anticipation for the extraordinary would also be the prelude to the working of the 'tourist gaze'.

What made the watering resorts an extraordinary realm for visitors was closely connected with their prestige, and the rise of excitement when approaching the tourist site was more easily brought to mind through paper peepshows than through topographical prints. The viewing of prints does not afford opportunities for building up anticipation, except perhaps in folio works by Lamb or Bruce where viewers developed their anticipation before they turned the page. Nonetheless, consisting only of text, the cover of the folio is purely functional and would give few indications of the extraordinary nature of the imageries inside. It was also difficult to simulate the sense of transition from the everyday realm into an unfamiliar world with the flipping of pages.

On the contrary, due to its structural features, the paper peepshow proves to be a suitable medium to evoke a sensation that shares similarities with that felt by tourists nearing their destinations. As many have pointed out, its cut-out panels make reference to the design of the theatre stage.[74] It becomes clear that the front-face (sometimes also the slipcase) has a similar function to the proscenium arches, as a boundary that signals the separation between users/audience and the scenes in the paper peepshow/

71 Borsay, 'A Room with a View', 185–6.

72 Susan Barton, 'General Introduction', in Susan Barton, ed., *Travel and Tourism in Britain, 1700–1914, vol. 1. Travel and Destinations* (London: Pickering & Chatto, 2014), ix.

73 Alison Byerly, *Are We There Yet?: Virtual Travel and Victorian Realism* (Ann Arbor, MI: University of Michigan Press, c.2013), 2.

74 Hyde, *Paper Peepshows*, 12–20.

onstage. This design is inherent to the paper peepshow as depictions on the cut-out panels belong to a different realm, just like the world onstage is a fictional one distinct from everyday life. As the gateway to the cut-out panels, the front-face hides the depictions from view and effectively raises curiosity and anticipation from users. Moreover, the process of opening the paper peepshow is also crucial. The moment of lifting the front-face constitutes an indispensable part of the joy of using this medium. This is a pleasure for both the visual and the haptic, as users took delight in seeing their manipulation enabling the previously hidden peep-view to slowly reveal itself and experience excitement and anticipation not unlike that associated with opening up a present. The affordance enabled by the paper peepshow structure also means that consumers could control the speed of their hands opening the works and enjoy the sensation at their own pace.

In paper peepshows depicting watering resorts, all these features could help with the realization of immediacy. The slow revelation of the peep-view generates emotions that could suggest those experienced on a journey. Since this feature is a part of the paper peepshow structure, it is present by default in all the works about watering resorts. However, publishers, including those of the works analysed here, often went further than using the existing structure and adopted various designs to effectively make use of characteristics of the paper peepshow in different ways to strive for a higher degree of immediacy. In *Interior View of Brighton Royal Chain Pier*, for example, a vignette on the slipcase depicts the Pier from a distance, from the sea, which is an unfamiliar perspective for users of the paper peepshow (See Figure 3.3). When users took out the work, they saw on the shutter a portrayal of the Pier from a familiar angle, from the entrance to the Pier, before they expanded the bellows and viewed the depiction of the Pier Head through the peephole. As users physically neared the portrayal of the Pier on the cut-out panels, they experienced the rise of anticipation for the final scene as they saw depictions that also gradually 'zoomed in' on the site, which is a sentiment that can resonate with the eagerness felt by tourists approaching the Pier itself.

In other paper peepshows, the emphasis of the design is mainly on the front-face. Its role as a boundary is stressed, which simultaneously gives the impression that the scenes depicted on the inside belong to a different

Figure 3.3: *Interior View of Brighton Royal Chain Pier*. Anonymous. Hand-coloured aquatint. 11 x 14 x 54 cm (expanded). *c.*1829. Front-face. Gestetner 215, Jacqueline and Jonathan Gestetner Collection, Victoria and Albert Museum, London. © Courtesy of the Victoria and Albert Museum, London.

realm, thereby contributing to the build-up of expectation. For instance, although the slipcases of the two *The Cheltenhamorama, a View of the Old Well Walk* also look quite plain, the front-face images of both works feature depictions of what appears to be a grotto or lush vegetation and evokes the users' curiosity to peer into the realm behind the peephole. This creation of anticipation was particularly successful with one of the front-face prints since the flag on the cut-out panel can already be seen through the peephole (See Figure 3.4). This glimpse of the inside of the paper peepshow could work like a teaser and further arouse the users' interest in the peep-view. In both cases of Brighton and Cheltenham, by evoking the excitement experienced by tourists more effectively than topographical prints, the paper peepshow can thus bring users closer to these towns. This effect, together with the portrayals of iconic sites – the Chain Pier

Figure 3.4: Cat 227 *The Cheltenhamorama, a View of the Old Well Walk*. Published by Henry Lamb. Hand-coloured lithograph. 15 x 10.8 x 69 cm (expanded). *c.*1832. Slipcase and Front-face. Gestetner 227, Jacqueline and Jonathan Gestetner Collection, Victoria and Albert Museum, London. © Courtesy of the Victoria and Albert Museum, London.

and Old Well Walk – that were known for their association with the fashionable society, could thereby also reinforce sentiments that highlight the non-everyday and, by extension, exclusive nature of these watering resorts.

Conclusion

During the early decades of the nineteenth century, when various print representations competed in their portrayal of domestic watering resorts, the paper peepshow might not have been a major player, but its significance should not be neglected. In focusing on the role as a commodity of this optical toy and the experience of consuming it, I have argued that the paper peepshow epitomizes how visualizations of a tourist site functioned to not only bring faraway scenes home but actively contribute to constructing people's perception of these places. As a non-material form of conspicuous consumption, going to domestic inland spas and seaside towns was a way for the English middle classes to display and confirm their status and wealth. Under the same category of behaviour was the purchase of certain non-utilitarian, semi-luxury goods, to which the paper peepshow belonged. Incorporating depictions of watering resorts in this print novelty could thus not only be a way for publishers to merge two symbols of taste and capital into one in their attempt to market their products. It was also a practice that reinforced the exclusive image of these tourist destinations through the role of the paper peepshow as a fancy article.

Drawing from the concepts of remediation and the tourist gaze, the comparison between paper peepshows and conventional topographical prints makes clear that the former can more effectively evoke the emotions and sentiments nineteenth-century middle classes experienced when approaching inland spas and seaside towns, an important part of which would be excitement of entering extraordinary places distinct from everyday life. With the development of the railway, from the 1840s watering resorts became increasingly less exclusive. Around the same time, the paper peepshow also lost its appeal as a fancy article. With their status-making nature fading, these towns and the paper peepshow were no longer the ideal combination. Indeed, after a Brighton paper peepshow was produced in 1842, inland spas and seaside towns ceased to be a subject matter for this optical toy, which continued to depict other types of tourist sites and landmarks, most famously the Great Exhibition of 1851. The sense of exclusiveness of watering resorts constructed in the paper peepshow did not stand the test of time and faded quickly, much like its ephemeral and fragile texture.

ANTHONY HAMBER AND STEVEN F. JOSEPH

4 Photographic Practice in Nineteenth-Century Tourism

The rise of topographical and landscape photography was closely linked to the development of tourism. Financial and social mobility acted as a catalyst, as did the introduction of 'holidays' from places of work. The affordability of travel by railway as a new form of transport and the associated rise in disposable income drove both the industrialization of tourism and leisure activities as well as the market for photographic prints and publications for tourists. Railways represented a primary driver. Passenger statistics support the sociological impact. On average, there were 0.65 railway journeys per head of British population in 1841, while in 1881 there were 20 per head of population.[1] As has been perceptively observed: 'Coincidentally, the train and camera were invented almost at the same time, and technological progress in photography paralleled the growth of railway lines in most developed countries, suggesting perhaps a symbiotic relationship.'[2]

Another stimulant for tourism was the development of interest in history, art, architecture and archaeology. Together with the opening of libraries, art galleries and museums on a municipal and local level, these interests spawned an associated range of publications, including tourist guides. John Urry has linked these two phenomena, suggesting that they

---

1 Dan Bogart, Leigh Shaw Taylor, and Xuesheng You, 'The development of the railway network in Britain 1825–1911' in *The Online Historical Atlas of Transport, Urbanization and Economic Development in England and Wales c.1680–1911*. Accessed on 1 August 2022 at: <https://www.geog.cam.ac.uk/research/projects/transport/onlineatlas/railways.pdf>.

2 Nuno de Avelar Pinheiro, 'Tourist Photography' in John Hannavy, ed., *Encyclopedia of Nineteenth-Century Photography* (New York and London: Routledge, 2008), 1398.

may constitute a self-reinforcing 'closed circle of representation' in which tourist photographs both reflect and inform destination images.[3] Another aspect is that of tourism as popularization or 'vulgarisation', such as suggested by the American novelist and critic Henry James (1843–1916) who observed that many of the Americans he saw in Europe in the late 1860s and 1870s were lacking all preparation for culture.[4]

Photographs documented the lifecycle of tourism. They could be collected as inspiration for travel at home or abroad, or the planning for future travel. They could be collected by tourists on their itineraries to their destination or on their return home. As the nineteenth century progressed, tourists could buy photographs in a variety of formats, ranging from individual prints, through portfolios of prints, albums and a range of photographically illustrated publications. The demographic of nineteenth-century tourists also determined the evolving market for photographs. Only a tiny minority of the Victorian working class secured, or were even in a position to seek, holidays with pay and, if requiring several consecutive days holiday in the summer months, needed the toleration if not the full approval of their employer. As the working class began to earn wages, this enabled them to divert expenditure into recreational channels; travel and its associated tourism industry grew to meet the demand. When the working class did spend money on photographs it was more likely to be on portraits of family, royalty, public figures and celebrities. Thus, the market for photographs targeting tourists was initially for the upper and rapidly expanding middle classes.

Italy represented a major tourist destination; between 5,000 and 10,000 British tourists visited Italy each year from the mid-1830s to the mid-1880s.[5] However, even 10,000 visitors in the 1880s represented a very

3 John Urry, 'The "Consumption" of Tourism', *Sociology*, 24/1, (1990), 23–35.
4 James Buzard, 'Ambivalent Appropriations: Culture and the Tourist' in James Buzard, *The Beaten Track: European Tourism, Literature, and the Ways to 'Culture', 1800–1918* (Oxford: Clarendon Press, 1993).
5 See Maria Antonella Pelizzari and Scott Wilcox, eds., *The Idea of Italy. Photography and the British Imagination, 1840–1900* (New Haven: Yale Center for British Art, 2022).

small proportion (0.033 per cent) of the 29.7 million population recorded in the 1881 UK census. Tourists visiting Rome could purchase daguerreotypes of topographical and architectural views from Lorenzo Suscipj (1802–85) as early as 1840.[6] Alexander John Ellis (1814–90), better known for his work in mathematics, phonetics, philology and music, was also an amateur photographer. Between 1840 and 1841 Ellis acquired or commissioned a group of 137 whole plate daguerreotypes (16.5 x 21.6 cm) of Italy, now held at the National Science and Media Museum in Bradford. Ellis purchased a number of daguerreotypes of Rome from Suscipj and Achille Morelli and aimed to use these to prepare etchings to be published in monthly instalments entitled *Italy Daguerreotyped: a collection of views, chiefly architectural, engraved after daguerreotypes in possession of the Editor*.[7] This publication never appeared.

Similar photographs found their way to the United Kingdom and into small local exhibitions. In October 1840, Mrs Amelia Opie (1769–1853), English novelist, leading abolitionist, and widow of the artist John Opie, R.A., exhibited a 'Daguerreotype view in Rome' at the exhibition of *Works of Art, Specimens of Natural History, Antiquities, Manufactures, Machinery and Models and Paintings, Sculpture, Engravings, &c, &c.*, held at the Norwich Mechanics Institution. The daguerreotype may have been acquired by her close friend Anna Gurney, who had been in Rome in April 1840. Commercial photographic views of foreign sites quickly found international distribution outlets. One of the earliest London outlets for photographs of Italy was Claudet & Houghton of High Holborn, London.[8]

John Pemble, *The Mediterranean Passion: Victorians and Edwardians in the South* (Oxford: Clarendon Press, 1987), 39–40.

6 Lorenzo Suscipj (1802–55), Italian optician and photographer operating from 182 via del Corso in Rome. His earliest daguerreotype is dated 8 February 1840.

7 Achille Morelli (1812–75), Italian photographer, active in Rome from 1840 or 1841. In June 1841 he took a 360-degree panorama of Rome from the Bell Tower on the Capitoline Hill using thirteen Daguerreotype plates.

8 In 1836, Antoine Claudet (1797–1867) and George Houghton (*c*.1805–87) formed a company that made various types of glass, including optical glass. The company expanded, increasing its range of photographic products. Claudet established his first daguerreotype photographic studio in London in 1841.

By mid-April 1840 the firm was advertising that it had daguerreotypes of Rome, Naples, Florence and Venice for sale.[9]

By 1854, following the upsurge in photography spawned by the 1851 Great Exhibition, booksellers were being encouraged to stock photographic images. An advertisement placed in *The Publishers' Circular* by Joseph Cundall (1818–75), who in 1852 had set up the commercial Photographic Institution in London, stated:

> PHOTOGRAPHS. – The Proprietors of the Photographic Institution are willing to forward to Country Book- or Printsellers, a Selection of Sixty Photographs, by the best English and Continental Artists, varying in price from Three to Twelve Shillings, on condition that at least Three Pounds worth are chosen, and the rest returned within a Fortnight. The usual Printsellers' discount is allowed – Address to Mr. Burningham, 168, New Bond Street.[10]

A decade later, in 1863, the London publisher and photograph dealer Alfred William Bennett (1833–1902) advertised photographs of Rome, Genoa, Verona, Venice and other continental cities, at 5s. each, including 'Views of the greatest interest to the Tourist, Architect, &c'.[11] The following year T. H. Gladwell, a London 'importer of Foreign Photographs', placed an advertisement in the *Alpine Guide Advertiser* that detailed some of the photographs in his stock and their geographic scope. (See Figure 4.1)

British visitors to Italy in the third quarter of the nineteenth century were steered by travel guides. The red covered handbooks of John Murray (1808–92) began publication in 1836 and became ubiquitous, so that *The Times* could observe in 1859 that the Englishman 'trusts to his *Murray* as he would trust to his razor, because it is thoroughly English and reliable'.[12] While the handbooks had significant detractors, such as John Ruskin (1819–1900) and Charles Dickens (1812–70), who both wrote in critical

9   *Athenæum* (18 April 1840), 305 and (5 December 1840), 953.
10   *The Publishers' Circular* (16 August 1854), 378.
11   *The Publishers' Circular* (1 July 1863), 332 (advertisement by A. W. Bennett).
12   *The Times* (22 September 1859), 10.

T. H. GLADWELL'S PUBLICATIONS

THE CHAIN OF MONT BLANC FROM THE FLEGERRE: a magnificent panorama, extending from the Col de Balme to Mount Joly, and showing the Brevent and Village of Chamounix, with Index to the Names and Altitudes of all the Peaks and Glaciers printed on the margin.

Price 16s. Also, Panoramas from the Brevent and the Buet

SWITZERLAND AND SAVOY. Four different Series, various sizes, comprising several hundreds of views of Grindelwald, Lauterbrunnen, Interlachen, Termatt [sic for Zermatt], Saas, Sixt, Zhun, Martigny, Chamouny, and the surrounding Peaks; also, Views of the Great Aletsch Glacier, unrivalled examples of Photography by F. Martens;

A series of 100 finely-executed tinted Lithographs, copies from the Swiss Photographs, price 2s. 6d. each

THE PYRANEES: A complete Collection of the exquisite Views of Maxwell Lyte.

ROMAN PHOTOGRAPHS: 500 different Views, including all the Ancient Monuments and principal Modern Buildings, copies of Antique Statues, Frescoes, &c. Price from 2s. 6d. to 30s. each.

NAPLES AND POMPEI: a series of 20 interesting Views. Price 3s. 6d. each.

In addition to the preceding, T. H. G.'s Collection comprises of Views in England, Scotland, Ireland, Wales, the Isle of Wight, France, Belgium, Germany, Austria, The Tyrol, Saxon Switzerland, the chief Italian Towns, Venice, Athens, Egypt, &c.

CATALOGUES OF ITALIAN and SWISS VIEWS may be had on application.

T. H. GLADWELL, Printseller, Publisher, and Importer of Foreign Photographs, 21 Gracechurch Street; and at the City Stereoscopic Depôt 87 Gracechurch Street, London, E.C.

Figure 4.1: Transcription of advertisement for 'T. H. Gladwell's Publications'.

terms about the series,[13] Murray's handbooks became 'the dominant textual lens through which English tourists experienced the country in the mid- to late nineteenth century'.[14] These guides recommended ex-patriot photographers, including Robert Macpherson (1811–72) and James Anderson (1813–77) in Rome, and John Brampton Philpot (1812–78) in Florence. Significantly, they also included advertisements of dealers in photographs placed in *Murray's Handbook Advertiser*, that by 1865 claimed to have an annual circulation of 15,000. One of these was Henry Wimmer's Gallery of Fine Arts in Munich, who acted as an agent for the shipping company J. & R. M'Cracken of London.

The routes to Italy taken by the British tourist varied. Some travelled across France and then along the Mediterranean coast. Others ventured via Belgium, down the valley of the Rhine and through Switzerland. One photographic publisher, E. Linde of Berlin, published around 1880 an extensive series of *carte de visite* views and monuments along the Rhine.[15] (See Figure 4.2)

How many outlets along the Rhine stocked *Bord du Rhin* has yet to be established. Linde, who had a branch office in London, also published stereoscopic views in a series titled *Der Rhein & seine Umgebungen*. Another publisher to target the tourist market was Adolphe Braun (1812–77) of Dornach in Alsace.[16] He produced series of views in stereoscopic

13   Peter Slater, 'Tourism, Perception and Genre: Imagining and Re-imagining Venice in Victorian Travel Writing', eSharp, *23: Myth and Nation* (Spring 2015), 1–15.

14   Stephanie Malia Hom, *The Beautiful Country. Tourism and the Impossible State of Destination Italy* (Toronto: University of Toronto Press, 2015), 43.

15   The firm of Ernst Linde, of Friedrichstrasse, Berlin, issued topographical photographic views from approximately 1862 to 1895, with its most active period between 1865 and 1880.

16   Adolphe Braun (1812–77), French photographer and publisher. In 1842, in Paris, he published a successful collection of floral designs. In 1847, he opened his own studio in Dornach, a suburb of Mulhouse. In the early 1850s he began to photograph studies of flowers. By the 1860s, he was publishing sets of photographs of the Alpine regions of France, Germany, Switzerland, and Italy. He also photographed in Belgium and Holland. He had a branch at 43, Avenue de l'Opera, Paris.

Figure 4.2: E. Linde. *130. Loreley. B. Vues du Rhin* Albumen Carte de visite.

Anthony Hamber collection.

format – such as *France, Suisse, Belgique*, and also *Bords du Rhin* – and built an international network of outlets.

Tourists also had an interest in the native costumes and customs of the localities they visited or passed through. This market was supplied by photographers creating series of portraits of locals in national or regional costumes and published in different photographic formats, such as *carte de visite* and cabinet cards. These were particularly popular on the Continent. Examples include Adolphe Braun's series *Costumes de Suisse* and Andries

Jager (1825–1905) of Amsterdam's *Costumes des Pays-Bas*.[17] In Italy Giorgio Conrad (1827–89) of Naples and Stefano Lais (1832–92) of Rome also published sets of costume studies that in some instances exploited prevailing cultural stereotypes.[18]

Hubert Jerningham (1842–1914) married Annie Mather (1850–1902) (née Liddell) in the Roman Catholic Church of the Assumption, Warwick Street, Westminster, on 3 December 1874. A 'Wedding Tour' photographic album provides a thoroughly documented itinerary for the couple's travel from London to Italy and back. It is significant since the album has a fully evidenced provenance for its creation and the photographs follow the itinerary of their honeymoon. Each photograph is captioned and dated. Hubert and Annie left for the Continent immediately after their wedding. They travelled from London to an English Channel port (probably Folkestone) and by boat to Boulogne sur Mer, then via Lyons, Marseilles, Nice, Monte Carlo (and, unusually, there are photographs of the interior of the dining room and gambling room of the Hôtel de Paris, opened in 1864, where they stayed) along the Corniche to Menton, San Remo, Ventimiglia, Genoa, Pisa, to Rome (of which there are many views, paintings, sculpture and architecture), where they had an audience with Pope Pius IX. The Jerninghams then travelled on to Florence, then Venice, and on to Munich. The last photographs are of paintings in the Louvre in Paris, after which it is assumed Annie and Hubert headed back to England.

A distinguishing feature of the large number of photographs in the Jerningham Album is that each photograph is captioned and annotated in manuscript. Critically, almost all the images have a date written below the right-hand corner. Thus, one has a calendar of the route of their honeymoon trip starting in Boulogne sur Mer on 4 December 1874 until the last image, dated 29 January 1875, a painting in the Louvre, 'Group of Artists'

17 Andries Jager (1825–1905), Danish-born bookseller, photographer, publisher and art dealer, who settled in Amsterdam and was active as a photographer from his studio at Het Water 110 from the late 1850s.

18 Giorgio Conrad (1827–89), Swiss-born photographer, active at Strada Fontana Medina 54 in Naples from the early 1860s. Stefano Lais (1832–92), Italian photographer, with a studio at 57 via di Campo Marzio in Rome from around 1860 until 1870.

by Velasquez. The last folio of the album has a poignant dedication, signed by both Hubert and Annie, and dated 31 January 1875. It reads:

> *All has an end in the world*
> *even*
> *the joy, pleasure, and delight of a wedding tour*
> *but what charming moments does an album recall!*

While photographs were expensive in the first two decades after their initial introduction, they became increasingly affordable. Establishing the relative prices of photographs across multiple nineteenth-century country currencies is problematic. However, the Murray handbooks did provide some details of prices, and even made subjective comments regarding these. In the 1857 edition of the entry for Venice there was no separate section for 'photographs' – or names of photographers – but an entry under 'Painters' that stated: 'The best views of Venice, and exquisite pictures of their kind, are the photographs, to be had at Munster's [i.e. Herman Munster "who speaks English"] and other printsellers in the Piazza San Marco, price four and eight zwanzigers each'.[19] In the same year, the cost of four Robert Macpherson photographs would have kept a tourist in Rome in lodgings in a private apartment with servants, including cook and housekeeper, for one week.[20] James Anderson was more aware of the expanding market for photographs. The 1862 Murray *Handbook to Rome* noted: 'Mr Anderson's photographs are also extremely faithful and good, and of different sizes to suit all purses and purchasers: they can only be procured at Spithover's library.'

While topographical studies, views of ancient monuments, and works of art (such as paintings and sculpture) proliferated, there are also examples of civic pride in progress, in the form of *carte de visite* photographs of newly completed municipal buildings, such as railway stations. Contemporary sources record that photographs purchased at commercial

---

19 *A Handbook for Travellers in Northern Italy Part 1 … Sardinia, Lombardy, and Venice* (London: John Murray, 1857), 310.

20 Mrs Steuart Erskine (ed.), *Anna Jameson Letters and Friendships* (London: T. Fisher Unwin, 1915), 304, 307.

outlets in Italy could be sent back to England using one of a number of shipping agents, one of the leading firms being J. & R. McCracken of London who had, by 1875 at the latest, an extensive network covering Italy.[21]

In 1865, the sexcentenary of Dante Aligheri (1265–1321) took place, celebrations that elevated him to the *de facto* Italian national poet, a neo-Ghibelline liberal who prophesied the unification of Italy.[22] In that year a number of statues depicting Dante were unveiled, notably one in Florence and another in Verona.[23] 'A Dundonian' visited Florence in 1868. He travelled on to Rome, where he visited the photographic establishment of Tommaso Cuccioni in via Condotti to buy photographic views. He wrote:

> We arranged to have all the photographs we had purchased in different places put together in a tin case, and among these was a photographic view of the statue of Dante [by Enrico Pazzi (1818–99)] standing in front of the Church of Santa Croce. The young man who was arranging them, on coming to this photograph, stopped and examined it, and called out 'Dante', and instantly everyone employed in the shop – masters, journeymen, and apprentices – assembled round him, and each one in his turn examined the photograph, and loudly expressed his admiration of the poet.[24]

There were a number of forms of tourism that exploited photography. One of these was battlefield tourism. Albums of photographs of historically significant locations were also produced, such as those recording the area

21 For instance, Murray's tourist handbooks provided such information on shipping agents as well as photographers.
22 The Guelphs and Ghibellines were rival parties in medieval Germany and Italy that supported the papal party and the Holy Roman emperors respectively. 'Neo-Ghibellines' thought the temporal power of the papacy a hindrance to Italian unification. See Mahnaz Yousefzadeh, 'Dante 1865 – The Politics and Limits of Aesthetic Education' in Joep Leerssen and Ann Rigney, eds., *Commemorating Writers in Nineteenth-Century Europe: Nation Building and Centenary Fever* (London: Palgrave Macmillan, 2014), 102–16.
23 *Inaugurazione del monumento a Dante Allighieri [sic] in Verona nel 14 maggio 1865* (Verona, 1865), with one albumen print of the statute by Ugo Zannoni (1836–1919).
24 *The Northern Warder and General Advertiser for the Counties of Fife, Perth and Forfar* (18 February 1868), 2.

in which the Battle of Waterloo took place in 1815. In 1854, one of the first British travel agents, Henry Gaze (1825–94), created a tour which included Waterloo. Nearly twenty years later, the Brussels-based photographer Alexandre de Blochouse (1821–1901) produced the first guide cum photographic souvenir to the battlefield, in English for its target audience of British visitors.[25] The publication is a handy size (15 x 11cm) to slip into a coat pocket, and consists of twelve albumen prints, bound in accordion style, that make up a circuit of the battlefield, accompanied by two pages of dense commentary written by retired army captain and firearms expert J. L. G. Charrin (1816–96). In the US there were similar battlefield examples, specifically those of the Civil War. William Henry Tipton (1850–1929) of Gettysburg described himself as 'The Battlefield Photographer'. By 1888, he claimed to possess over 5,000 battlefield views and 100,000 portrait negatives. He produced souvenir albums and photographs in a variety of sizes, presumably to match the pockets of visiting tourists.

Parlour entertainment formed another environment for the conjoining of tourism and photography. From the 1860s, the hiring of lantern projectors and accompanying slides of foreign sites developed into a popular type of entertainment in formal and informal venues. By 1865, the 'Wholesale Photographic Warehouse' of the Salisbury photographer James Miell (1837–1901) advertised that it offered 'photographic views of Egypt, the Holy Land, Rome, London, &c., &c., lanterns and slides on sale or hire'.[26] Another relevant example of the way in which photographs covered the entire lifecycle of travelling tours can be found in the late nineteenth-century activities of Toynbee Hall, created in 1884 in the Whitechapel district of the East End of London with the aim of sharing knowledge and culture amongst the local men and women of 'humbler social status'. Overseas tours were seen as a way of furthering their cultural education. In August 1887, eight 'Toynbee pilgrims' visited the principal towns of Belgium, and the Toynbee Travellers' Club was created

<hr>

25 *Waterloo (Views and Text) published by the Belgian Royal Society of Photographs* (Brussels–Ixelles, 1873). Accessed on 1 August 2022 at: <https://www.sydney.edu.au/museums/collections/photography/waterloo.shtml>.

26 *Salisbury Journal* (8 April 1865), 5.

in 1889. Italy formed the destinations of the Club's tours in its earliest years: Florence (1888), Venice (1889), Siena, Perugia, and Assisi (1890) and Florence again (1891). The Club formed photographic albums from these tours, and this built on an 1888 appeal by the Education Committee for the donation of photographs.

Extensive preparation for these Toynbee Hall tours took place, such as for the 1888 tour: 'Italy was our first goal, and so during the winter of 1887 the would-be travellers heard lectures, saw photographs, and read books on Italy and her history, Florence and her art, Milan and her galleries, and Antwerp and her buildings.'[27] The Toynbee Travellers acquired photographs while on the tours, both through purchases, and taken by amateur photographers in their ranks, some of whom owned Kodak No. 1 cameras, invented and marketed by George Eastman (1854–1932), and introduced in 1888. The Toynbee Travellers were said to have returned from their Venice tour in 1889 with around 2,000 photographs.[28] On the 'travellers' return from an excursion, a selection of the travel photos was collected in albums, displayed at Toynbee Hall, and then added to its growing collection of photographs.

The cathedral city of Salisbury offers an example of a typical British Victorian provincial city in which to examine the rise and impact of photography on tourism.[29] In 1839, the year of the announcement of photography, Salisbury was a comparatively small market city. In the 1841 census there were 10,086 inhabitants recorded. The city was also a resting-place for the stagecoach from London to Exeter, and Bristol to Southampton. It was estimated that in 1839, 37,000 passengers travelled through Salisbury on the Exeter to London stagecoach.[30] The expansion of the British railway network during the 1840s brought significant benefits to Salisbury. The London and Southwest Railway (LSWR) opened its Milford station

27 Henrietta Barnett, *Canon Barnett: his life, work, and friends*. Vol. I (London: John Murray, 1918), 412 footnote.
28 Marcella Pellegrino Sutcliffe, 'The Toynbee Travellers Club and the Transnational Education of Citizens, 1888–90', *History Workshop Journal*, 76 (2013), 137–59.
29 See Anthony Hamber, *The Origins of Photography in Salisbury 1839–1880* (East Knoyle: Hobnob Press, 2019).
30 John Chandler, *Endless Street – A History of Salisbury and Its People* (Salisbury: Hobnob Press, 1983), 137.

on the east side of the city on 1 March 1847 and in the following year the rail connection between Salisbury and London was completed via Bishopstoke. The impact of the expanding railway network on Salisbury was noted in 1867 at the opening ceremony of the purpose-built eponymous museum of the wealthy financier William Blackmore (1827–78) located behind the Salisbury and South Wilts Museum in St Ann Street. Horatio Bolton Nelson, 3rd Earl Nelson (1823–1913), pointed to Salisbury's emerging position as a transport hub, stating that the city was 'so generally connected by railways with almost every other place in England' that the establishment of the Blackmore Museum in the city 'might really be called a boon to the nation at large'.[31]

The Gothic Revival had taken off around 1840. Salisbury had seen the reception of architect Augustus Welby Northmore Pugin (1812–52) into the Catholic Church in 1835 and in 1848 his Roman Catholic church of St Osmund was completed in Exeter Street. The interest in the Gothic coupled with the increased number of visitors to the city resulted in Salisbury Cathedral becoming one of the most photographed buildings in Great Britain. The building was to help sustain many of Salisbury's commercial photographers and draw a number of leading contemporary photographers to document the building, including Roger Fenton (1819–69).[32] In 1840, William Henry Fox Talbot (1800–77), one of the inventors of photography, had been appointed High Sheriff of the County of Wiltshire and travelled to Salisbury to sit in the Assizes court that was held there on a quarterly basis each year. Talbot was encouraged twice by his uncle William Thomas Horner Fox Strangways, 4th Earl Ilchester (1795–1865), career diplomat, botanist, and art collector, to photograph the cathedral. On 7 March 1840 Strangways wrote a letter to Talbot providing detailed consideration of viewpoints and technical issues. Then on

---

31 *Some Account of the Blackmore Museum – Part I. The opening Meeting* (Devizes: H. F. & E. Bull; London: Bell & Daldy; London: J. R. Smith, 1868), 35. See also Anthony Hamber, *Collecting the American West. The Rise and Fall of William Blackmore* (East Knoyle: Hobnob Press, 2010).

32 For the significance of these photographs by Fenton, see Anthony Hamber, 'Photography and the West Front in the mid-Nineteenth Century' in Tim Ayers, ed., *Salisbury Cathedral. The West Front* (Chichester: Phillimore, 2000), 131–8.

2 November 1841 Strangways wrote from Frankfurt, thanking Talbot for 'your beautiful Calotypes' and suggesting 'I wish some fine day you would take the cathedral of Salisbury.'[33] Talbot appears to have never followed up on his uncle's suggestions.

However, Talbot was aware of the demand from the rapidly evolving market for travellers. In a letter to Henry James (1803–77), the Director of the Ordnance Survey in Southampton, dated 25 May 1859, Talbot makes it clear that he could envisage commercial opportunities for making copies of 'special selected localities' from larger or general maps [of the Ordnance Survey] and reproduced by his photoglyphic engraving photomechanical process:

> The process [photoglyphic engraving] is so easy that there would be no difficulty for example in mapping the environs of any or all the principal towns in England. If these were collated into a volume with an Index, it would form an atlas of a novel kind, useful to the traveller.[34]

The *Salisbury Journal* noted in September 1847 that specimen daguerreo-types by Barber, of James's Terrace in Winchester, could be seen at Brodie and Co., in the Canal, perhaps indicating that the existing booksellers and print sellers in Salisbury stocked or could source photographs during this decade. In September 1853 William Russell Sedgfield (1826–1902) was advertising in the *Salisbury & Winchester Journal* that a set of ten photographs of Salisbury could be obtained from Messrs Brown & Co. or F. A. Blake. Queen Victoria visited Salisbury on 15 August 1856 en route from Plymouth to Osborne on the Isle of Wight and it was reported that Messrs. Brown had supplied her with a copy of Dodsworth's *Salisbury Cathedral*, with India proof impressions, and also several photographs of the cathedral and other views of Salisbury by Sedgfield, one of which was a view of the High Street.

---

33 The letters are in the Fox Talbot Museum in Lacock. *The Correspondence of William Henry Fox Talbot* <https://foxtalbot.dmu.ac.uk/> accessed on 1 August 2022. Document numbers 04056 and 04353.

34 *The Correspondence of William Henry Fox Talbot,* <https://foxtalbot.dmu.ac.uk/> accessed on 1 August 2022. Document number 7891.

George Brown and his son William – who died in February 1867 aged 35 – built their business in the Salisbury's Canal into a veritable emporium, selling and publishing books, and offering book binding, stationery, prints, and photographs. In the 1857 edition of *Brown's Stranger's Handbook and Illustrated Guide to the City of Salisbury* there is a full-page advertisement by Brown & Co. that includes 'Photographs, by Fenton, Sedgfield, and others – a great variety, from 10*s*. 6*d*. downwards. Stereographs of the Cathedral, Streets of Salisbury, Poultry Cross, Wilton, Stonehenge, Old Sarum, &c., by Sedgfield, Wilson, &c., 1*s*. and 1*s*. 6*d*. each'. This evidences that Brown was an early provincial stockist of photographs by leading British photographers. By the 1860s, photographs by non-Salisbury photographers could be acquired from a number of outlets in Salisbury, including bookshops, print sellers, and local photographers.

While Brown's handbooks to Salisbury, and to the cathedral, consistently included advertisements for his stock of prints and photographs placed on the inside of the front and back cover, the company never published photographically illustrated editions, always preferring wood engraved illustrations. These advertisements included Francis Frith's *Universal Series*, the 1876 catalogue for which numbered over 4,000 photographs classified under domestic and foreign locations. Frith developed an international network to purchase images for, and then to distribute his *Universal Series*. For instance, they could be purchased through E. & H. T. Anthony of The Broadway, New York, who advertised 'Beautiful 8 x 10 [inch] photos of all parts of the world'.[35] Further research is required to establish the scale and scope of the local, regional, national, and international distribution networks for photographs.

Photographs for tourists came in a variety of formats. Loose prints were mass-produced and targeted a broad market. The *carte de visite* and stereo view, popular from the late 1850s, are the earliest examples, followed by the cabinet card in the late 1860s. (See Figure 4.3)

Alongside single prints, a variety of photographically illustrated publications catered to the tourist trade. They were of various levels of

---

35　*Frank Leslie's Illustrated Newspaper* (14 October 1871), 80.

Figure 4.3: Anonymous photographer. Twelve Views of Salisbury Cathedral
[a sampler]. Carte de visite. Albumen print with the label of Westley's Library,
Cheltenham on the verso. Anthony Hamber collection.

sophistication and were widely divergent in presentation. The common factor was the process used, at least until the 1870s and often beyond: albumen prints, produced and mounted individually.[36] At the simple end of the publishing spectrum were viewbooks, generally small-format sets of topographical photographs that initially competed with series of lithographed or steel-engraved views. The photographs were usually in the standard formats of *carte de visite* or cabinet cards.[37] Viewbooks were often cloth-bound with a gilt title on the front cover and bereft of interior letterpress. They were generally produced locally and in short print-runs by a stationer or photographer (sometimes one and the same person), laid out in accordion style or bound, and often comprised diverse issues with a differing number of prints. Such publications were rarely recorded bibliographically and must count as orphans in printing history.

Francis Bedford (1815–94) began his career as an architectural draughtsman and lithographer, before taking up photography in 1851 or 1852. He became one of the most successful photographers of topographical views of the Victorian era, and he specifically exploited the tourist market by means of the publication of viewbooks. Bedford's photographs adhered closely to the sites and routes recommended in tourist guidebooks. He aligned with contemporary issues relating to cultural tourism and also health tourism.[38] Over ten per cent of Bedford's published output consisted of Welsh subjects, perhaps surprising since he was based in London. From around 1856 to 1880 Bedford took some 900 negatives of Welsh views, primarily in the 'wild' northern part of the country. In addition, Bedford came to a publishing arrangement with a firm in Chester,

---

36 The standard bibliography is Helmut Gernsheim, *Incunabula of British Photographic Literature 1839–1875* (London & Berkeley: Scolar Press, 1984). It contains 638 entries for publications, including periodicals, until 1875. This total considerably underreports the actual number of titles produced, due partly to Gernsheim's exclusive criteria for inclusion and partly to the pre-internet era in which his research was conducted.

37 Mounted on card stock measuring approximately 6.5 x 10.5cm and 10.5 x 16.5cm respectively.

38 See Stephanie Spencer, *Francis Bedford, Landscape Photography and Nineteenth-Century British Culture. The Artist as Entrepreneur* (Farnham: Ashgate, 2011).

Catherall & Prichard, that in 1860 commenced publishing his stereo views as a series entitled *Chester and North Wales Illustrated*. Around 1865 Bedford began a series of viewbook publications under the imprint of Catherall & Prichard. During the 1860s and 1870s the firm issued around eighteen photographically illustrated albums of topographical views.[39] These included views of Wales and West Country seaside resorts.

Viewbooks, such as Bedford's, could also be issued in series where flexibility of presentation was the byword for commercial longevity and for appealing to a broad range of pockets, bound in either gilt-decorated morocco or cloth, and several in both octavo and quarto versions. Bedford's basic template per title was an oblong octavo with ten albumen prints captioned in letterpress but copies are known with varying totals, the number ranging from a minimum of ten prints to sixteen, twenty, twenty-four and to a maximum of thirty. The title page was standard to each locality or area covered and titles omitted the date of publication and a table of contents. In June 1866 Catherall & Prichard retired and the business was purchased by Philipson & Golder.[40] However, titles continued to be released under the Catherall & Prichard imprint, undoubtedly in small runs and possibly using pre-printed generic title pages, thereby enabling the publishers to issue copies on an ad hoc basis and Bedford to update and substitute any of the views to be included.

More sophisticated products than viewbooks, photographically illustrated guidebooks, combining images and text, began to be published in numbers from the 1860s onwards. When properly conceived, the text would offer a structured itinerary whilst the accompanying photographs

39 Pictorial Illustrations of Torquay and its Neighbourhood. A Series of Photographic Views; Photographic Views of Warwickshire; Photographic Views of Bristol and Clifton; Photographic Views of Exeter; Photographic Views of Ilfracombe; Photographic Views of Chester; Photographic Views of Beddgelert; Photographic Views of Devonshire; Photographic Views of North Devonshire; Photographic Views of South Devonshire; Photographic Views of Stratford-on-Avon and Neighbourhood; Photographic Views of Tenby and Neighbourhood; Photographic Views of Dartmouth and the Dart; Photographic Views of Shrewsbury and its Neighbourhood; Photographic Views of Bettws y Coed; Photographic Views of North Wales; Photographic Views of Torquay. Gernsheim, Incunabula, 74–5, lists thirteen titles, giving an 1865–75 range of publishing dates.

40 *Chester Courant* (27 June 1866), 1.

would add value to the tourist's experience. The Buchan railway, in the north-east of Scotland, was built to link the fishing ports of Fraserburgh and Peterhead with Aberdeen. On completion of the Fraserburgh stage in 1865, William Anderson, editor of *The Peterhead Sentinel*, published a cloth-bound guidebook to attractions along the line.[41] It is typical of the first generation of photographically illustrated publications in this category. It was published alternatively in an unillustrated and a 'photographic edition'.[42] The latter contained nine albumen prints *hors texte*, taken by a local rather than a nationally known photographer, Joseph Collier (1836–1910) of Peterhead. The photographs were meant to complement the text in a publication aimed to accompany the tourist travelling from station to station 'along the route' as the guide's subtitle indicates. Thus, a view of Pitfour House is inserted facing a commentary that opens: 'A little to the right [of the line] we now pass the grounds of Pitfour. From the line we do not see the house, which is a magnificent structure, fronted by large Corinthian pillars. The grounds, however, are the greatest attraction to those who might care to spend an hour or two in the locality.'[43] The photographer was prominently acknowledged, with a printed credit on each plate and in an advertisement at the rear that included the phrase 'Landscape photography especially attended to, and on Moderate Terms'.

Perhaps Collier was too successful since in March 1866 he sold his studio, together with his stock of glass negatives, to James Shivas (1836–1930). Consequently, later re-issues of Anderson's guidebook credit Shivas as the photographer rather than Collier, demonstrating the tenuous nature of copyright and authorship in early photography. In fact, many guidebooks mis-attributed or simply failed to acknowledge the photographer, even when photographic illustrations were a publication's unique selling point. *A Concise Guide to Maidstone*, undated but published in 1883, is

---

41 William Anderson, *The Howes o' Buchan, being notes, local, historical, and antiquarian, regarding the various places of interest along the route of the Buchan railway* (Edinburgh: W. P. Nimmo; Aberdeen: Lewis Smith; Peterhead: 'Sentinel' Office, 1865).

42 Not in Gernsheim, *Incunabula*. We have only located copies of the photographic edition.

43 Anderson, *The Howes o' Buchan*, 58.

typical in this respect. As stated on the cover, this unpretentious oblong octavo publication, in stiff printed boards and cloth backstrip, was priced at *1s* and illustrated 'with five photographic views'. The printer-publisher J. Burgiss-Brown was keen to emphasize the contribution of the photographs to a publication imbued with civic pride: 'The Illustrations, more costly than those usually employed for the purpose, he also hopes may be acceptable. To issue at the popular "Shilling" has been an important feature with this edition.'[44] The photographer's identity is passed over in silence. The advertising section at the rear confirms that Burgiss-Brown was a dealer in photographs ('Views of Maidstone in several sizes') and albums ('Half-guinea Portrait Album, the best in the trade') so he supplied the illustrations himself, but nothing indicates whether he was also the photographer.

There is one category of photographic publication that was a high-Victorian phenomenon: volumes of verse or novels illustrated with site-specific photography in Romantic mode. The London publisher Alfred W. Bennett, specializing in 'photographic gift books', was quick to see the appeal of such publications.[45] Bennett and his successor, Provost & Co., published several titles of this type in the 1860s and early 1870s, focusing their efforts on the work of two of the most popular and successful authors of the nineteenth century, William Wordsworth (1770–1850) and Walter Scott (1771–1832). The two poets' evocation of natural landscape found huge resonance amongst a readership that hankered after an antidote to industrialization and the urban environment, a target market for literary tourism.

Wordsworth was a successful bard of the heritage landscape and object of secular pilgrimage. Readers of his Romantic poetry wanted to see for themselves the settings in the Lake District that he had immortalized. Bennett not only tailored his literary output to a receptive readership but

44 [J. Burgiss-Brown], *A Concise Guide to Maidstone* (Maidstone: J. Burgiss-Brown, [1883]), 'Preface', unpaginated.

45 For a biography and checklist of publications, see Steven F. Joseph, 'Alfred W. Bennett and the Photographic Gift Book', *The Rijksmuseum Bulletin*, 70/3 (2022), 223–43.

set down a plausible itinerary for potential, returning or armchair tourists. *Our English Lakes, Mountains, and Waterfalls as Seen by William Wordsworth* went through four editions in six years, initially for the 1863–4 Christmas gift season. The anthology opens with an immediately recognizable tourist shrine to the author, a frontispiece view of Wordsworth's home. (See Figure 4.4)

The ensuing verse and accompanying albumen prints by specialist landscape photographer Thomas Ogle (1813–82) are structured by location explicitly for the active tourist, an approach that Bennett explained in an introduction to the first edition and reprinted in subsequent editions:

> [The Compiler] has, as far as practicable, classified these extracts under the heads of the different Lakes or other objects of interest in each locality. By this arrangement it is believed that not only will the Reader be able, with the assistance of the Photographic Illustrations, ... to appreciate the more fully Wordsworth's wonderfully true descriptions of the beauties of Nature; but the Tourist will have the additional pleasure of identifying with his own favourite spot ...[46]

Walter Scott was William Wordsworth's opposite number in Scotland as an object of literary veneration. Geoffrey Wakeman has summarized the appeal of photography inspired by Scott's works thus: 'Photographs made ideal souvenirs and some books were obviously aimed at this market. Queen Victoria's passion made it fashionable for tourists, and publishers capitalized on the fact.'[47] In Bennett's and Provost's case, this meant the re-publication of four of Scott's verse narratives set in a medieval or chivalric past, repackaged as photographic gift books, beginning in 1863 with the acclaimed poem *The Lady of the Lake*, followed in 1866 by *Marmion: A Tale of Flodden Field*, then *The Lord of the Isles* in 1871 and finally *The Lay of the Last Minstrel* in 1872. For a title-page vignette to *Marmion*,

---

46 Our English Lakes, Mountains, and Waterfalls as Seen by William Wordsworth (London: A. W. Bennett, 1868 (third edition), v–vi.

47 Geoffrey Wakeman, Victorian Book Illustration: The Technical Revolution (Newton Abbott: David & Charles, 1973), 'Chapter Five: The Sixties – The Impact of Photography', 85.

Figure 4.4: Thomas Ogle, 'Rydal Mount, the Residence of Wordsworth', albumen print, 9.0 x 8.5cm, frontispiece to Our English Lakes, Mountains, and Waterfalls as Seen by William Wordsworth. (London: A. W. Bennett, 1864).

(Courtesy of Rijksmuseum, Amsterdam, RP-F-2001-7-361-1).

Bennett featured not the expected landscape but a study by Thomas Annan (1829–87) of the author's monument in Edinburgh, a popular starting point of the Scott trail.[48]

This form of cultural commodification, the marketing of literary texts complemented by evocative photography with tourism as an underlining *raison d'être*, reached its apogee with the replacement of the usual ornamental cloth or morocco binding with varnished wooden boards decorated with transfer illustrations or in tartan. The application of Mauchline ware, named after the East Ayrshire town which was the centre of the trade, was not confined to photographically illustrated books but other publications tended to be decorated in a more restrained form with simple transfer prints.[49] On the other hand, Scott's photographically illustrated editions of the 1860s and 1870s would typically bear albumen print cartouches on the upper and lower boards, smaller transfer prints on the corners of the upper board and often a statement of authentication such as 'made from wood grown on the land of Abbotsford' (Scott's estate) or 'made from wood grown on Flodden field' in the case of *Marmion*. Furthermore, a set of metal or mother-of-pearl studs would be affixed to the lower boards, thereby transforming a work of literature into a hybrid artefact, a *livre-objet* suitable for display in many a high-Victorian drawing room.

Tourists to Italy were exposed to a similar initiative for commodifying literature, where their individual experience of travel could be recast in the light of journeys in fiction. Travel reminiscences could be reinforced by the purchase of photographically customized copies of works with Italian subject matter, mainly fiction, bound in gilt-decorated vellum or mock vellum and extra-illustrated with topographical views and reproductions of art works relevant both to the text at hand and the tourist's specific

---

48 For a discussion of mid-Victorian tourist photography in Scotland, see Roger Taylor, *George Washington Wilson Artist & Photographer (1823–93)* (Aberdeen: Aberdeen University Press, 1981), 'Chapter Five: Tourism and Photography', 51–60.

49 David Trachtenberg, and Thomas Keith, *Mauchline Ware: A Collector's Guide* (Woodbridge: Antique Collectors Club, 2002), 'Books Published in Mauchline Ware Boards', 266–73.

itinerary. These books are often associated with Bernhard Tauchnitz (1816–95), the Leipzig publisher, whose reprint of *The Marble Faun* by Nathaniel Hawthorne was an English-language bestseller on the tourist trail in Rome and consequently the work most often encountered in this *deluxe* extra-illustrated form.[50] However, it was the local booksellers or print dealers, rather than Tauchnitz, who created this new product, came to agreements with professional photographers, and enabled customers to make their own selection of photographs which were then often mounted into pre-bound copies on leaves left blank for this very purpose.[51]

Although novels under the Tauchnitz imprint constitute the best-known examples of these artefacts for tourists, travel books and art history also feature as well as fiction from other publishers. Internal evidence suggests that the practice flourished from the 1870s until after 1900. The focus in the descriptive opening of *The Improvisatore*, a Bildungsroman by Hans Christian Andersen first published in 1835, cannot fail to contrast with the facing photograph of a contemporary square captured in suspended motion. (see Figure 4.5)

> Whoever has been in Rome is well acquainted with the Piazza Barberini, in the great square, with the beautiful fountain, where the Triton empties the spouting conch-shell, from which the water springs upwards many feet. Whoever has not been there, knows it, at all events, from copperplate engravings.[52]

The unintentional dissonance of text and image, with the fountain, focus of the opening lines, relegated to a corner of the photograph, plus the incidental reference to an older form of graphic art – so many elements that

---

50 Ann Wilsher, 'The Tauchnitz *Marble Faun*', *History of Photography*, 4/1 (1980), 61–6; Paul Edwards, 'Musing with the muse in the photographically illustrated Marble Faun', Word & Image, 30/1 (2014), 64–75.

51 Victoria Mills, 'Vision, History, and the Tauchnitz Editions of George Eliot's Romola', in Maria Antonella Pelizarri, and Scott Wilcox, eds., *The Idea of Italy: Photography and the British Imagination 1840–1900* (New Haven, CT: Yale Center for British Art, 2022), 96–101.

52 Hans Christian Andersen, *The Improvisatore, or, Life in Italy* (London: Ward, Lock & Co., n.d.), 1. The dedication in the copy consulted is dated Rome, 13 March 1907.

Figure 4.5: Anonymous, [Piazza Barberini], silver gelatin print, 8.0 x 12.1cm,
in Hans Christian Andersen, *The Improvisatore; or, Life in Italy*
(London: Ward, Lock & Co., Limited, before 1907).

(Courtesy of Rijksmuseum, Amsterdam, RP-F-2001-7-638).

remind us that in the relationship between tourism and visual production
in the nineteenth century, commercial photography, while a burgeoning
beneficiary, never quite had the last word. In this instance, as in many
others, and whether intended for the 'discerning' traveller or the mass vaca-
tioner, photography was revealing its two faces: a multiform and popular
consumer durable and a personalized pattern of visual mnemonic.

5 Two Devon Tourists of 1863: The Illustrated Travel Journals of Harriette Armytage and John Follett

This chapter brings together two manuscript travel journals illustrated with the steel line-engraved vignette views that were popular in the mid-nineteenth century. The first was compiled by Harriette Armytage who travelled from Yorkshire to Torquay, Devon in 1863–4 and the second was written by John Follett of Topsham, Devon who travelled to the Lake District in 1863. Both shed interesting light on the distribution and use of these precursors of the picture postcard.

In the Westcountry Studies Library in Exeter are three small albums bearing the title 'Drives &c in and about Torquay', by H. M. A. The journal is contained in three sketch books and includes eighty-three illustrations, including two maps. Volume 1 covers from 8 October 1863 to 26 November 1863, volume 2 from 27 November 1863 to 12 April 1864 and volume 3 from 13 April 1864 to 2 May 1864.[1] The bookseller had not provided information about their provenance but a little investigation and more than a little luck identified H. M. A. as Harriette Matilda Armytage, born 11 July 1843 and baptised 26 August 1843 at Hartshead-cum-Clifton, Yorkshire. She was the daughter of Sir George Armytage, 5th Baronet (1819–99) who had married Eliza Matilda Radcliffe at Spofforth on 1 June 1841. The family was notable in the parish, endowing a school there, and the local inn being named the Armytage Arms. Their seat was Kirklees

---

1 Harriette Matilda Armytage, 'Drives &c in and about Torquay', Manuscript, 1863–4. Held by Westcountry Studies Library, Devon Heritage Centre. <https://etched-on-devons-memory.blogspot.com/2017/03/drives-in-and-about-torquay-by-h.html> accessed 2 May 2022. Images of individual pages can be accessed at this URL.

Hall, a grade one listed building in Hartshead, now a set of gated private apartments near junction 25 of the M62.[2]

Tourism was an important driver of printing innovation and, even before the Armytage family arrived in Torquay on 8 October 1863, they had had encounters with various forms of print. They travelled by train via London, so would have required railway timetables, a relatively new typographical format, but one mastered by printers many years before the family set out on their journey, as Bradshaw's 1859 timetable for the Devon line shows. The timetable layout, still familiar today, had been foreshadowed by the development of statistical and mathematical tables, particularly since the introduction of the census in 1801.[3]

Their next encounter with print would be the railway ticket. The laborious process of hand-writing tickets had been replaced in 1842 following the introduction of the Railway Clearing House and the development of the Edmondson railway ticket, a system for recording the payment of railway fares and accounting for the revenue raised. Tickets were now printed on card cut to $1\,\frac{7}{32}$ x $2\,\frac{1}{4}$ inches (31.0 x 57.2 mm) and they were stored in standard ticket racks with each ticket in a series individually numbered as early surviving Devon railway tickets show.[4]

The family arrived in Torquay by train from London, 'by the G[reat]. W[estern]. R[ailway]. and S[outh]. D[evon]. R[ailway]. having passed several fine views also the towns of Reading, Bath, Bristol, Exeter &c. A very fine view may be seen before the train arrives at Newton and leaves the River Teign for Torquay. Shortly before the traveller arrives at the station he sees the fine avenues of Torre Abbey on the left'.[5]

2 Sir Bernard Burke, *A Genealogical and Heraldic History of the Peerage and Baronetage*, 76th edition (London: Harrison & Sons, 1914), 124–6. <https://archive.org/details/bub_gb_Pf8cAAAAYAAJ/> accessed 2 May 2022.

3 *Bradshaw's General Railway and Steam Navigation Guide for Great Britain and Ireland*, Issue 306, (London: W. J. Adams, 1859), 33. <https://www.google.co.uk/books/edition/Bradshaw_s_General_Railway_and_Steam_Nav/HSApAAAAYAAJ>

4 Wikipedia, 'Edmondson railway ticket', <https://en.wikipedia.org/wiki/Edmondson_railway_ticket> accessed 2 May 2022.

5 Armytage, vol. 1, 2.

They had rooms in Cumper's Private Hotel 'in Sulyard Terrace, the first house coming from the station in the Terrace'. It was located on the seafront, facing towards the harbour. On arrival there was an important visit to be made to number 10 The Strand, the premises of E. Cockrem the printer and publisher of the *Torquay directory and South Devon journal*.[6] In common with many newspapers in health resorts, this weekly carried a regularly updated directory of the principal residents and visitors. New arrivals would be distinguished with a special mark in the listings, and it was essential that such an important man as Harriette's Papa should make his special mark among the other visitors.

During this visit Harriette probably purchased the sketch books to be used for her journal, two maps of Torquay which she tipped into the front of the first and second volumes of her journal, and Cockrem's newly published guidebook to Torquay and its neighbourhood, from which she quotes. She also must have started her collection of engraved vignettes early in her stay and provided herself with pen and ink and a supply of paste and a brush.[7] The two maps also show Cockrem gearing up to meet the requirements of tourists. The *Plan of the Town of Torquay from the Ordnance Survey with corrections and additions to the present time, being a companion to the Torquay Directory* was lithographed by G. D. Palmer of Exeter and published by Edward Cockrem, 10, Strand, Torquay, in about 1863. To each side it has a street index, naming individual houses, thus enabling the purchaser to locate other visitors they may wish to meet. This was at the large scale of eight inches to a mile.[8]

The other, *Map of Torquay and the adjacent country, within a distance of 8 miles* was drawn by Edward Appleton of Torquay, lithographed by William Spreat of Exeter, and published by Edward Cockrem, bookseller & stationer, Directory Office, Torquay, and was dated 1863. This map, at the smaller scale of two inches to a mile was used to navigate the family

6 Armytage, vol. 1, 11.
7 *Cockrem's tourists' guide to Torquay and its neighbourhood* (Torquay: E. L. Seeley, 1864).
8 Armytage vol. 1, map.

on their drives in and around Torquay.⁹ On 13 October they took possession of Kanescombe, a villa in Lower Warberry Road rented by her 'Papa'
Harriette was delighted:

> The view on the South west side of the house is very beautiful, overlooking the
> garden (in which there are several pretty trees, shrubs, plants &c) and through the
> valley between the hill called the Braddons and Park Hill there is also a good view
> of the Bay, the village of Paignton on the opposite shore, and a beautiful range of
> hills beyond, which stretches right & left.¹⁰

Papa worked hard at making a mark. He was a passionate horserace breeder,
so was invited to be steward of the Torquay steeple chases together with
members of the county gentry, and this is mentioned in Exeter newspapers,
but COVID-19 meant that it was impossible to check local files of the
Torquay newspapers for the period of Harriette's stay and these are not
yet covered online by the British Newspaper Archive. On 7 November
the family attended a concert at the Bath Saloon with an international
cast of performers including Carlotta Patti, sister of the famous singer
Adelina Patti from Madrid, and others from Germany and La Scala,
Milan. Unfortunately, she remarked, the piano was not working properly
that evening. There are few other mentions of their various social activities, but we know Miss Armytage was at an assembly with her parents on
22 January 1864.¹¹

It is also impossible to say exactly who were in the party. Apart from
Papa and Mama (Eliza) it is probable that Harriette's younger brothers
Arthur Henry (born 1845) and Reginald Francis (born 1849) were with
them at least during the holidays. On Saturday 9 January 1864 Harriette
writes: 'Drove to Park Hill on the Torbay Road and to the Railway Station
to meet the boys after the paper chase'.¹² Papa was sometimes called away;
Harriette reports going to meet him at the station on 20 November, but

---

9   Armytage vol. 2, map.
10 Armytage vol. 1, 3.
11 Armytage vol. 1, 23.
12 Armytage, vol. 2, 7.

it would seem that she was largely left with her mother to her drives, rides, walks and collection of vignette views. She seems to have been an observer rather than a participant in Torquay's social life, but quite a keen observer as a lively pencil sketch of Torquay scandal mongers shows.[13]

Harriette had started writing up her diary before 24 October 1863, the date of a correction she made, so she had collected many of the engravings by that date. They were neatly pasted in ones or twos at the foot of each page, leaving room for the text above. The images often bear some relation to the text, which was mainly added later.[14] The bulk of the illustrations are of Torquay, but there are four of Dawlish, three of Teignmouth, two of Berry Head and Brixham and one each of Devonport, Berry Pomeroy, the mouth of the Dart, Compton Castle and the Teign Estuary.

Harriette did not venture far on her drives and there was inevitably much revisiting of favourite locations on the 120 drives and rides that she took when weather permitted. Her furthest point south was Goodrington, a distance of about six miles from Kanescombe. To the north she reached as far as Maidencombe, about four miles away and to the east Cockington, about three miles distant. In all she pasted in eighty-one engravings, mostly small steel line-engraved vignettes around 60 x 95mm in size, but there was also a selection of a dozen larger vignette views about 110 x 180 mm in size, some in the larger format series published by Besley in Exeter. They come from a mix of London and local publishers, showing the range of items available in booksellers' shops at that time.[15]

13   Armytage, vol. 3, 30.

14   Armytage, vol. 1, 12.

15   Ian Maxted, *The Devon Book Trades: a Biographical Dictionary*, (Exeter: J. Maxted, 1991). This gives biographical details of most Devon book and print trade members. It can be supplemented by: Ian Maxted, 'The Devon Book Trades Directory: Torquay', <https://bookhistory.blogspot.com/2014/07/devon-book-trades-torquay.html>, accessed 3 May 2022, for Devon practitioners and: Ian Maxted, 'Biographical dictionary of Devon Printmakers', (2017) <https://etched-on-devons-memory.blogspot.com/2017/03/biographical-dictionary.html>, accessed 3 May 2022, which also includes London printmakers.

Steel line-engraved vignettes began to be published in the 1840s, stimulated by the spread of the railways and the penny post. They were used to illustrate guidebooks or were published in booklets of six, twelve, eighteen or more views. Separately they were issued as letterheads or sometimes with two views together on a sheet of notepaper. They were also issued separately on card or paper, and it was in this form that Harriette acquired them. Those available to her in 1863 originated with about ten publishers. Of the eighty-one engraved vignettes collected by Harriette twenty-three were not recorded by John Somers Cocks who in 1977 published a catalogue of more than 3,502 topographical prints published between 1660 and 1870, both separately published and also included in 229 different series. It included many hundreds of steel line-engraved vignettes.[16]

J. or J. & F. Harwood seem to have been the first to issue numbered series of engraved vignettes, starting about 1841 in 26 Fenchurch Street, London and they continued to at least 1856. Normally they give the publisher's name to the left and the exact day of publication and number to the right. There are nineteen examples in Harriette's journal.[17] Kershaw & Son, also working in London, did not normally date their prints but they seem to have been working between about 1845 and 1860. There are seven examples in the journal.[18] William Frederick Rock was the leading London publisher of vignette views, producing some 7,000 nationwide between 1846 and 1878. He was a Barnstaple man and became a major benefactor to his native town, so it is perhaps not surprising that Devon is well represented among his vignette views. The earliest Devon ones appeared, without numbers, as illustrations to Rachel Evans *Home scenes, or Tavistock and its vicinity* in 1846, but after that his prints are carefully numbered and dated. With twenty-four examples in the journal, Rock is the best represented of all the publishers, accounting for more than a

---

16 J. V. Somers Cocks, *Devon Topographical Prints 1660–1870* (Exeter: Devon Library Services, 1977).

17 Armytage, vol 1, frontispiece, 6, 7, 26, 29, 31, 33, 35, 36; vol. 2, frontispiece, 1, 8, 16, 23, 25; vol. 3, frontispiece, 1, 5, 25.

18 Armytage, vol 1, 8, 9, 22, 24, 25; vol. 2, 6, 27.

quarter of the total.[19] Another series of numbered vignettes was also issued from London signed J. S. & Co. They were undated but were probably published in the period around 1855–65. The publisher has been identified as J. Simmons who continued to be active in the 1870s. There are five examples in Harriette's journal.[20]

Among local publishers Henry Besley, with fifteen vignettes in Harriette's journal, is the best represented. Working in Exeter he produced a larger series covering Devon and Cornwall, mainly after the local artist George Townsend. They are unnumbered but normally dated and sixty-six views of Devon appeared between 1848 and 1871. In 1853 he started a smaller series, and these are numbered, the Exeter series starting at 100, earlier numbers being reserved for Cornwall. They were used to illustrate his *Route books* of Devon and were also issued in small booklets with titles such as *Peeps at Exeter and neighbourhood*.[21]

Torquay publishers provide a meagre haul. Edward Croydon with six vignettes is the best represented. He was a printer, stationer, bookbinder, librarian, and music seller at 2, Victoria Parade from 1840 to 1878. He was the son of Edward Croydon, bookseller in Teignmouth where he was born in 1816. His premises was known as the Royal Library, and it figures thus in imprints of local guides in the early 1860s. At least one of his engravings was made for him by John Newman & Co. of London.[22] Daniel Ardley, represented by four prints, was a bookseller and stationer in Braddons Row, recorded in directories between 1848 and 1856. Some of his publications were engraved by Rock & Co. in London. He may have ceased trading before Harriette appeared on the scene, but his vignettes were still in stock.[23]

19 Armytage, vol. 1, 2, 5, 16 (2), 17 (2), 18 (2), 30, 32, 37; vol. 2, 2, 3, 5, 18, 19, 21; vol. 3, 6, 10, 12, 15, 17, 27, 28.
20 Armytage, vol. 1, 6, 14, 27, 30, 34.
21 Armytage, vol. 1, 4, 10, 11, 12, 28; vol. 2, 4, 10, 11, 12, 14, 29; vol. 3, s 8, 13, 16, 18.
22 Armytage, vol. 1, 3, 5, 15, 19, 29 (engraved by Newman & Co., London); vol. 3, 14 (published in Teignmouth)
23 Armytage, vol. 1, 1, 13, 37 (engraved by Rock & Co., London); vol. 2, 3 (engraved by Rock & Co.), 9.

The other main bookseller in Torquay was Edward Cockrem who had set up his office as Cockrem & Elliott in about 1834. He had been apprenticed to John Hannaford of Totnes and was a printer, bookseller, stationer, music seller, and ran a circulating library at 10, The Strand and imprints are recorded up to his death in 1872. He also had premises in Bridge Street, Newton Abbot in 1855. Harriette only picked up one of his vignettes, but her two maps were both published by him. The vignette, which was engraved for him by Rock & Co., is one of the few from outside Torquay and, showing the train puffing along beside the Teign Estuary, it may have served as a souvenir of her journey down to Torquay. Cockrem published the *Torquay and Tor directory* newspaper from 1839 until at least 1856 and in 1848 was agent for the Society for Promoting Christian Knowledge (SPCK).[24] There was also one stray vignette published by William Wood of Devonport pasted into Harriette's journal.[25]

Torquay had an active book and printing trade in the early 1860s, although it is not well recorded.[26] William Elliott was a printer, bookseller and stationer who also ran a library and reading rooms. Apprenticed to Lazarus Congdon of Plymouth Dock, he began business as Cockrem & Elliott about 1834. He separated from Cockrem and moved to Vaughan Parade about 1840, moving in the early 1850s to 2, Lawrence Place where he is listed as proprietor of the *Torquay Chronicle* in 1856. William Fisher was active as a bookseller and publisher from 1863 to 1883 based at 14, Fleet Street between 1878 and 1883. John Robinson, a printer is recorded at 10, Lower Union Street in 1856, 86, Lower Union Street in 1878 and Cotford Terrace, Upton Hill in 1893. He published the *Torquay Recorder* from 1861 to 1867. Henry Powlson was a printer with several publications to his name between 1861 and 1864. E. L. Seeley also published a number of items between 1860 and 1877 including an edition of *Cockrem's tourists' guide to Torquay and its neighbourhood*.

24 Armytage, vol. 1, map, 2 (engraved by Rock & Co.); vol. 2, map.
25 Armytage, vol. 3, 26.
26 Ian Maxted, 'The Devon Book Trades Directory: Torquay' (2014). <https://bookhistory.blogspot.com/2014/07/devon-book-trades-torquay.html>, accessed 3 May 2022.

There were two or perhaps three newspapers published in the town in 1862 and 1863, the *Torquay and Tor directory* had been established in 1839 and by 1863 it had changed its name to the *Torquay directory and South Devon journal.* As already mentioned, its proprietor was Edward Cockrem. The *Torquay Chronicle* ran from 1849 to 1865 with William Elliott as its proprietor and the *Torquay Recorder*, proprietor John Robinson appeared between 1861 and 1867 although no copies of this title are known to survive.[27] There were also local artists to produce prints. John William Salter (1825–91), artist and lithographer was active from the 1850s. He produced large lithographs and paintings of the Torquay area.

However, the age of the engraved vignette was nearing its end. Already the first booklets of photographs were appearing, and the new technical wizardry of the stereoscopic photograph was soon to drive engraved vignettes from the scene. From 1863 Francis Bedford paid annual visits to Devon and the sequence of his negative numbers implies that he landed in Ilfracombe each year and made his way south, reaching Torquay in the first year of his visits. He was preceded by Francis Frith, whose earliest Devon photographs date from 1861, although Torquay appears only to have been visited around 1870. There are no photographs in Harriette's journals, although she should have been able to acquire examples during her stay.[28]

William Rock issued his last engraved vignette in 1876. He sold the firm of Rock Brothers in 1884 and, among other benefactions to Barnstaple established the North Devon Athenaeum in Barnstaple whose library holds albums of several thousand of his vignettes. As for Harriette, she neatly rounded off the account of her drives with indexes, both by page and alphabetical, and left with her family for London on 2 May 1864. She died the following year aged twenty-two and was buried on 13 October 1865 at Hartshead. She was clearly one of the many consumptives who sought

<hr>

27 Jean Rowles and Ian Maxted, *Bibliography of British Newspapers: Cornwall* [and] *Devon*, (London: British Library, 1991).
28 Ian Maxted, 'Early Publishers of Devon photographs: Francis Bedford', (2018) <https://devon-bibliography.blogspot.com/2018/10/devon-bibliography-photographs-bedford.html>, accessed 3 May 2022.

relief in Torquay, and the drives were undertaken to give her plenty of fresh air. There is no evidence that the family returned to Torquay the following winter.

Although such illustrated journals rarely survive, Harriette was not the only person in Devon to compile such a travel account in 1863. Four months before she set out on her journey to Devon, John Follett, of Mount Wear, Topsham, a substantial residence overlooking the Exe Estuary, had completed his own illustrated manuscript entitled 'Notes of a Tour in the Lake District, May 1863'. It is a single octavo volume of 173 numbered pages, plus six pages of preliminary matter with thirty-seven engraved views, bound in blue cloth, and gilt. It covers a period of three weeks from 27 April to 18 May 1863 and includes Wasdale Head, Windermere 'till lately called Birthwaite', Rayrigg House 'formerly the residence of Wilberforce', Bowness, Hawkshead, Coniston, Grasmere, Rydal, Sty Head Pass and other localities. It mentions Wordsworth's house at Rydal, Harriet Martineau's house at Ambleside, Southey's monument in Crosthwaite Church and meeting with Edgar Nembhard Thwaites (1836–1919), curate of St John's Keswick.[29]

It provides a detailed account of a tour of the Lake District, with many cameos of their experiences, for example Wasdale Head, 'which consists of only a few … houses, the principal one of which to visitors is the first approached, being that of John Ritson, where refreshments for man & horse can be obtained [...] kindly waited on by the two pretty, sweetly mannered [...] daughters' or Sty Head Pass, where 'we mounted our ponies under the protection of a guide, a most active and intelligent old fellow of the name of Ben'[30]

29 John Follett, 'Notes of a Tour in the Lake District, May 1863', (Manuscript, 1863). In private collection. I am greatly indebted to Timothy Sykes who drew my attention to this item in 2021 at the conference 'A visitor attraction: printing for tourists' and has been able to examine this volume and other Lakeland publications of the period to provide background information for this section of the paper.
30 Follett, 141–2, 136.

The manuscript volume seems to have reached the open market for the first time at a sale at the Exeter auctioneers Bearnes Hampton & Littlewood as lot 339 of the sale held 23 March 2016. The present owner intends that it will be added to the collections of a local library or museum in the Lake District. The writer, John Follett (1806–66), was the brother of William Webb Follett (1796–1845), for some time the nation's Attorney General. He was a timber and hemp merchant, who lived for some years in Mount Wear House, Topsham, Devon. His daughter Laura E. Follett, born 1841, inscribed her name in the front cover with the date Feby. 8 [18]80.[31] The layout of the 'Tour' is meticulous. Where an engraving appears (usually in the upper half of the leaf in the manner of notepaper) text is written below but never on the verso of the sheet co-extensive with the engraving on the recto. This means the illustrations are seen to best advantage.

One conjectures whether the 'Tour' was compiled in its final form while en route. Alternatively, the sheets of notepaper may only have been acquired while in the Lake District with the intention of preparing a fair copy on return utilizing the sheets of notepaper. Certainly, the illustrations are carefully selected to be relevant to the adjacent text perhaps indicating a retrospective work.

The landscape of engraved vignettes in the Lake District differed from that of Devon. The London publishers were represented, but only J. & F. Harwood seem to have published booklets of views in the 1840s and 1850s. Rock & Co. never seem to have ventured that far north-west. Their place was filled largely by the Edinburgh firm of William Banks & Co. Most of the engravings in 'Notes of a Tour in the Lake District' appear to be taken from *Views of the English Lakes on notepaper*. The Wordsworth Trust holds an example of this, which was 'engraved by W[illiam]. Banks and Co.; drawn by Theophilus Lindsay Aspland' (1807–90). It was published in Windermere by John Garnett, probably sometime in the 1850s. The engraved front cover gives some indication of the coverage:

31 R. E. Wilson, 'The Folletts', *Devon and Cornwall notes and queries*, 34 (1979–81), 94–7, 149–54.

VIEWS OF THE ENGLISH LAKES ON NOTE PAPER / WINDERMERE / WINDERMERE FROM LOWOOD / RYDAL WATER / GRASMERE / VALE OF KESWICK / DERWENT WATER / CONISTON / WAST WATER / BORROWDALE / HONISTER CRAG / ARA FORCE / ULLSWATER / Engraved by W. Banks, Edinburgh.[32]

The illustrations used by John Garnett in Windermere for his sales of novelty illustrated writing paper also appeared in Garnett's own publications. The most common of these are in oblong octavo format under the title *Views of the English Lakes*, containing captioned illustrations only but no text. Such volumes contain typically twenty-four or thirty plates with no great uniformity of content between the various issues. One assumes Garnett had a wide-ranging stock to draw on. One copy of such *Views* in the private collection of the owner of the 'Tour' has a manuscript inscription dated 'May 1860', very much a contemporary of Follett's travel account.

In those books, almost all plates are 'Engraved by W Banks Edinburgh' with or without the addition of '& Son'. In addition, many are shown as 'Drawn and Engraved' by Banks. Some, but not a great number, however, are shown as 'L. Aspland Delt' with Banks's name also recorded. What thus appears in Garnett's published books is very much reflected in his issues of notepaper used in the 'Tour'. William Banks & Sons advertisements in other versions of Garnett's *Views of the English Lakes* in the same private collection describe the myriad forms of printing issued by Banks including 'Headed Letter Papers' and 'Letter Tops'.

William Banks (1806–66) was a skilled draughtsman as well as an engraver and most of the views are from his own drawings, with others by Theophilus Lindsey Aspland (1807–90), a pupil of George Cooke. Banks's relationship with the publisher John Garnett was close enough for him to name one of his sons John Garnett Banks (1860–90). John Garnett (1825–96) was a prominent figure in Windermere.[33] He established his printing

<hr>

32 William Banks & Co., 'Views of the English Lakes on notepaper' (Windermere: John Garnett, 1850/59).

33 List of Garnett's imprints on JISC: <https://discover.libraryhub.jisc.ac.uk/search?publisher=Garnett&publisher-place=Windermere>, accessed 23 May 2022.

business by 1852 when together with the London publisher Whittaker & Co. he produced *Keswick and its neighbourhood: a hand-book for the use of visitors, to all the scenery, nooks, and corners of the district, with a unique map, shewing the carriage roads, usual sailing route round the lake, and the foot-paths open to the public.* In 1854 his *Guide to Windermere* by Harriet Martineau was illustrated with engravings after T. L. Aspland, as was *A complete guide to the English Lakes*, also by Harriet Martineau, which was published by John Garnett in Windermere, and Whittaker & Co. in London, 1855. In a directory section this publication lists 'Garnett, John, railway superintendent, – Printer, bookseller, &c., Post-Office'. The railway had reached Windermere from Kendal in 1847 and from 1853 he published *Garnett's railway time tables and guide to all public conveyances in the Lakes District*, which continued under a variety of names and with extended coverage until 1896.[34] Later in his career he published some of the works of William Wordsworth, including an edition of *The excursion: a poem* in about 1860, an anthology of his poems in 1874 and, many years after Hudson and Nicholson of Kendal had ceased to issue the work, *A description of the scenery of the district of the Lakes*, which probably appeared as late as 1886.

Apart from a single illustration of Windermere from J. Harwood, 26 Fenchurch Street, London, no. 228 in his nationwide series, there are two engravings by the Scottish engraver and painter William Home Lizars (1788–1859) which were published by John Hudson, who was active in Kendal as a printer, bookseller, publisher, stationer and newsagent from 1825 to 1858, the date of his last imprint. He was also partner with Cornelius Nicholson in the Cowan Head Paper Mills to the north of the town. Thomas B. Hudson, perhaps his son, operated briefly in Kendal in 1859, and some of his headed notepaper was still available when John Follett arrived in the Lake District four years later.[35]

---

34 Some issues have been digitized: Garnett's railway time tables and guide to public conveyances (Windermere: J. Garnett, August 1868–July 1870), <https://www.google.co.uk/books/edition/Garnett_s_railway_time_tables_and_guide/CoEGAAAAQAAJ>, accessed 23 May 2022.

35 List of Hudson's imprints on JISC: <https://discover.libraryhub.jisc.ac.uk/search?publisher=Hudson&publisher-place=Kendal>, accessed 23 May 2022.

Brought together by chance, these two documents show two very different ways that Devon tourists in the early 1860s used the cheap steel line-engraved vignettes to illustrate their travels, both of them individuals in the more affluent ranks of society, one a young lady travelling south for health reasons and keeping a scrapbook journal day by day pasting the individual prints into sketch books, the other a gentleman of mature years travelling north as a sightseeing tourist and carefully arranging the illustrated notepaper to receive a manuscript account written after his return and then to be bound up for him. The collections survive because they have been assembled into books; many others must have been dispersed because they remained as loose sheets. Both shed interesting light on the range of engraved images available to tourists at this time, and further research to locate and compare other similar collections would be valuable.

Table 5.1: Table of prints included in the Armytage and Follett manuscripts

The Somers Cocks numbers refer to J. V. Somers Cocks, *Devon topographical prints 1660–1870: a catalogue and guide* (Exeter: Devon Library Services, 1977).

Armytage Vol: page	Title / artist, engraver	Publisher	Date	Series number	Somers Cocks
A1:002	Town & Harbour, Torquay	London: J. & F. Harwood, 26, Fenchurch St	[1842?]		3192
A1:003	Map of Torquay and the adjacent country, within a distance of 8 miles	Published by Edward Cockrem, bookseller & stationer, Directory Office, Torquay	1863		[Map]
A1:01	Waldon Hill & Sulyard Terrace	A. Cowan & Sons, Cannon St West. Published by D. Ardley	[1855?]	No. 447	Not SC

Table 5.1: (Continued)

Armytage Vol: page	Title / artist, engraver	Publisher	Date	Series number	Somers Cocks
A1:02	The River Teign, from Brecknock Hill, Devon	Rock & Co., London. Published by E. Cockrem, Newton Abbot & Torquay	10th March 1855	No. 2696	Not SC
A1:03	Torquay	Published by E. Croydon Public Library	[1855?]		Not SC
A1:04	Public Gardens, Torquay. / G. Townsend del. Exeter	Published by H. Besley	Pubd. April 4th 1853	No. 116	3326
A1:05	Babbacombe	Published at Croydon's Royal Library	[1855?]		Not SC
A1:05	Watcombe, near Torquay	Rock & Co. London	6 Aug. 1860	No. 1786	3363 [1852]
A1:06	Torquay from Stantiford Hill	London J. Harwood, 26, Fenchurch St	25 April 1856	No. 296	3318
A1:06	Torquay from Park Hill No. 2	J. S. & Co.	[1860?]	No. 1784	3242
A1:07	Torre Abbey	London, J. Harwood, 26, Fenchurch St	Feby 12th 1845		3350
A1:08	The Beacon, Torquay	Kershaw & Son London	[1850?]	No. 512	Not SC
A1:09	The Bason & Pier Head, Torquay	Kershaw & Son London	[1850?]	No. 516	3199
A1:10	Torquay from Corban Head. / G. Townsend del.	Pub. By H. Besley, Exeter	[1861?]	No. 159	3298

(Continued)

Table 5.1: Table of prints included in the Armytage and Follett manuscripts
(Continued)

Armytage Vol: page	Title / artist, engraver	Publisher	Date	Series number	Somers Cocks
A1:11	The Strand, Torquay. / G. Townsend del. Exeter	Published by H. Besley	Pubd April 4 1853	No. 113	3082
A1:12	Tor Church. / G. Townsend del. Exeter	Published by H. Besley	[1853]	No. 114	3130
A1:13	Cockington Church, near Torquay	Published by D. Ardley	[1855?]		Not SC
A1:14	Babbacombe Bay from the Beacon	J. S. & Co.	[1860?]	No. 1789	3062
A1:15	Watcombe, Torquay	Pub. By E. Croydon, Royal Library, Torquay	[1855?]		Not SC
A1:16	Torquay, from Park Hill (No. 1)	Rock & Co. London	1st Feby 1851	No. 1543	3236
A1:16	Torquay, from Park Hill (No. 2)	Rock & Co. London	1st Feby 1851	No. 1544	3237
A1:17	Torquay, from Park Hill (No. 3)	Rock & Co. London	April 10th 1851	No. 1611	3238
A1:17	Torquay, from Park Hill (No. 4)	Rock & Co. London	April 10th 1851	No. 1612	3239
A1:18	Daddy-hole, Torquay, Devon	Rock & Co. London	2 Aug. 1860	No. 1535	3158 [1851]
A1:18	Hesketh Crescent, Torquay	Rock & Co. London	29 Sep. 1860	No. 1138	Not SC
A1:19	Ansty's Cove, and Bishopstow, Torquay	Pub. By E. Croydon, Royal Library, Torquay	[1855?]		3021
A1:22	Torquay from the Abbey Sands	Kershaw & Son, London	[1860?]	No. 851	3296

Table 5.1: (Continued)

Armytage Vol: page	Title / artist, engraver	Publisher	Date	Series number	Somers Cocks
A1:24	Torquay, from the Pier	Kershaw & Son, London	[1845?]	No. 517	3193
A1:25	Torquay, from Livermead	Kershaw & Son, London	[1860?]	No. 1163	3297
A1:26	Chapel Hill, Torquay. / Wm Redaway	London, J. Harwood, 26, Fenchurch Street	June 9 1846		3323
A1:27	Torquay from Park Hill. No. 1	J. S. & Co.	[1860?]	No. 1783	3241
A1:28	The New Church Torquay. / G. Townsend del. Exeter	Published by H. Besley	[1853?]	No. 115	3143 [variant title?]
A1:29	Hesketh Crescent, from the Sea. / Eng. By Newman & Co., 48 Watling St London	Pub. At Croydon's Royal Library & Reading Rooms, Torquay	[1848?]		3076
A1:29	The Cove under Daddy's Hole, Torquay	London, J. Harwood, 26, Fenchurch St	22 June 1854	No. 360	3161
A1:30	Torquay from the beacon	Rock & Co. London	13 June 1860	No. 1143	Not SC
A1:30	Park Hill from the beacon	J. S. & Co.	[1860?]	No. 1780	3240
A1:31	Tor Abbey House, Torquay	London, J. Harwood, 26, Fenchurch Street	Augt. 6th 1847	No. 144	Not SC
A1:32	Bishopstowe, South Devon	Rock & Co. London	Jany 16th 1851	No. 1524	3105
A1:33	Daddys Hole, Torquay	London J. Harwood 26 Fenchurch St	June 10th 1854	No. 312	Not SC

(Continued)

Table 5.1: Table of prints included in the Armytage and Follett manuscripts
(Continued)

Armytage Vol: page	Title / artist, engraver	Publisher	Date	Series number	Somers Cocks
A1:34	Waldon Hill, Torquay, No. 2	J. S. & Co.	[1855?]	No. 1642	3295
A1:35	The Strand, Victoria Parade, Park Hill, &c. Torquay	London J. Harwood 26 Fenchurch St	21 April 1854	No. 283	3083
A1:36	Entrance to the Basin, Tor Abbey, Waldron Hill	London J. Harwood 26 Fenchurch St	May 25 1854	No. 291	Not SC
A1:37	Chapel Hill Torquay	Rock & Co. London. Published by D Ardley	Dec 1st 1850	No. 1490	3324
A2:002	Torquay with Waldrom Hill & Pier. / C M Redavy	London J. Harwood 26 Fenchurch St	Jany 1st 1848	No. 744	3197
A2:003	Plan of the Town of Torquay from the Ordnance Survey with corrections and additions to the present time, being a companion to the Torquay Directory. / G. D. Palmer lith. Exeter. Scale 8 inches to a mile	Published by Edward Cockrem, 10, Strand, Torquay	[1863?]		
A2:01	Town Hall & Independent Chapel Torquay	London J. Harwood 26 Fenchurch St	July 27 1854	No. 333	Not SC
A2:02	Natural Arch, Torquay	Rock & Co. London	14 Aug 1860	No. 1491	3164 [earlier]
A2:03	Abbey Crescent, Torquay	Rock & Co. London. Published by D Ardley	1 Nov 1861	No. 4413	3069

Table 5.1: (Continued)

Armytage Vol: page	Title / artist, engraver	Publisher	Date	Series number	Somers Cocks
A2:04	Babbicombe from Petit Tor. G. Townsend del	Published by H. Besley, Exeter	April 4th 1851	No. 157	3063
A2:05	Cricket Ground, Torquay, Devon	Rock & Co. London	15 Apr 1855	No. 2737	Not SC
A2:06	Torquay	Kershaw & Son	[1845?]	No. 511	3305
A2:08	Berry Head & Brixham	London J. Harwood 26 Fenchurch St	June 1 1846	No. 695	0250
A2:09	Torr Church, Devon	Pub. By D. Ardley Torquay	[1855?]		Not SC
A2:10	The Baths, Berry Head &c. Torquay	Pubd by H. Besley	April 4th 1853 [?]		3178
A2:11	Babbacombe	Published by Henry Besley, Exeter	[1855?]		Not SC
A2:12	Watcombe. – near Torquay. / G. Townsend del. Exeter	Published by H. Besley Directory Office, South Street, Exeter	[1855?]		3365
A2:14	Meadfoot	Published by H. Besley Directory Office, South Street, Exeter	[1855?]		3151 [Variant title?]
A2:16	Natural Arch on the Road to Upton, Torquay	London J. Harwood 26 Fenchurch St	Jany 1st 1848	No. 745	Not SC
A2:18	Torquay, from Waldon Hill	Rock & Co. London	24 June 1863	No. 4723	3274
A2:19	Torquay, from Park Hill	Rock & Co. London	24 June 1863	No. 4724	3252
A2:21	Torre Abbey, South Devon. / Fuller del.	Rock & Co. London	Feby 1st 1850	No. 1308	Not SC

(Continued)

Table 5.1: Table of prints included in the Armytage and Follett manuscripts
(Continued)

Armytage Vol: page	Title / artist, engraver	Publisher	Date	Series number	Somers Cocks
A2:23	Bishopstowe	London J. Harwood 26 Fenchurch Street	April 4th 1845		3013
A2:25	Torquay, from Tor Abbey Lawn	London J. Harwood 26 Fenchurch Street	July 20th 1846	No. 707	3287
A2:27	The Bathing Cove, Torquay	Kershaw & Son London	[1850?]	No. 515	Not SC
A2:29	Daddy's Hole	Published by H. Besley Directory Office South St Exeter	[1855?]		3159?
A3:002	Town & Harbour, Torquay	London J. & F. Harwood 26 Fenchurch St	[1842?]		3192?
A3:01	Entrance to the Basin, Tor Abbey Waldron Hill, Torquay	London J. Harwood, 26 Fenchurch St	May 23 1854?	No. 281? or 291?	Not SC
A3:02	Torquay from the Rock Walk	[Manuscript title, print cropped]	[1860?]		Not SC
A3:05	Park Hill, Torquay	London J. & F. Harwood 26 Fenchurch Street	[1842?]		3267
A3:06	Babbacombe, Devon	Rock & Co. London	[1859?]	No. 1012	3058
A3:08	Ansty's Cove, near Torquay	Published by H. Besley Directory Office, South St Exeter	[1855?]		3020

Table 5.1: (Continued)

Armytage Vol: page	Title / artist, engraver	Publisher	Date	Series number	Somers Cocks
A3:10	Torquay from the Pier Head	Rock & Co. London	8 June 1860	No. 1144	3198 [1849]
A3:12	Teignmouth, Devon. / Fuller delt	Rock & Co. London	Novr. 1st 1849	No. 1241	2929 [1849]
A3:13	Teignmouth Beach. / G. Townsend del.	Published by H. Besley	[1853]	No. 109	2866
A3:14	Teignmouth Devon	Library Teignmouth	[1845?]	No. 909	2926A?
A3:15	Dawlish, Devon	Rock & Co. London	[1848?]	No. 402	0585
A3:16	Dawlish Lawn. / G. Townsend del. Exeter	Published. By H. Besley	[1853]	No. 106	0587
A3:17	Dawlish, Devon	Rock & Co. London	[1848?]	No. 400	0616
A3:18	Luscombe, Dawlish. / G. Townsend del. Exeter	Published by H. Besley	[1853]	No. 107	0650
A3:25	Berry Pomeroy Castle Devonshire	London J. & F. Harwood 26 Fenchurch Street	Nov.16th 1842		0145
A3:26	Devonport to Torpoint. / Sketched by W. Hake	Pub. By W. Wood, Devonport	[1860?]		1979
A3:27	Mouth of the Dart	Rock & Co. London	1 Sep 1858	No. 3801	Not SC
A3:28	Compton Castle, South Devon. / Fuller del.	Rock & Co. London	Feby. 1st 1850	No. 1309	Not SC

(Continued)

Table 5.1: Table of prints included in the Armytage and Follett manuscripts
(Continued)

Follett	Title / Artist, engraver	Publisher	Date	Series number
F00	The Knoll	[Listed in contents but not included]		
F01	Derwent Water / Engraved by W Banks Edin	[Windermere: J. Garnett?],	[1854/59]	
F003	Windermere from near the Hotel / Drawn & Engraved by W Banks Edin.	Windermere: J Garnett	[1854/59]	
F007	Windermere from the Crown Hotel	London: J Harwood 26 Fenchurch Street	[1854/59]	No. 228
F013	Bowness from Belle Isle Windermere / Drawn & Engraved by W Banks and Son Edin.	[Windermere: J. Garnett?],	[1854/59]	
F015	The Ferry Windermere / drawn & Engraved by W Banks and Son Edin	[Windermere: J. Garnett?],	[1854/59]	
F017	Windermere Lake (Storrs Hall) / Drawn & Engraved by W Banks and Son Edin	[Windermere: J. Garnett?],	[1854/59]	
F019	Esthwaite Water / Drawn & Engraved by W Banks and Son Edin	[Windermere: J. Garnett?],	[1854/59]	
F025	Coniston / L. Aspland delt; W. Banks sc, Edin.	[Windermere: J. Garnett?],	[1854/59]	
F027	Windermere from Lowwood Hotel / Drawn & Engraved by W Banks Edin	[Windermere: J. Garnett?],	[1854/59]	
F031	Wray Castle Windermere / Drawn & Engraved by W Banks and Son Edin	[Windermere: J. Garnett?],	[1854/59]	

Table 5.1: (Continued)

Follett	Title / Artist, engraver	Publisher	Date	Series number
F041	Stock Ghyll Force Ambleside / Drawn & Engraved by W Banks and Son Edin	[Windermere: J. Garnett?],	[1854/59]	
F047	Loughrigg Tarn Engraved Banks & Son Edin	[Windermere: J. Garnett?],	[1854/59]	
F049	Grasmere from Red Bank / L. Aspland delt; W. Banks sc, Edin.	[Windermere: J. Garnett?],	[1854/59]	
F051	Prince of Wales Lake Hotel Grasmere / Engraved Banks & Son Edin.; Photographed from the Island For J Garnett Windermere	Windermere: J. Garnett,	[1854/59]	
F059	Colwith Force / W Banks & Son Edin	[Windermere: J. Garnett?]	[1854/59]	
F065	Dungeon Ghyll / Eng by W Banks & Son Edin	[Windermere: J. Garnett?]	[1854/59]	
F069	Lower Fall at Rydal / L. Aspland delt; W. Banks sc, Edin.	[Windermere: J. Garnett?]	[1854/59]	
F071	Upper Fall at Rydal / Engraved by W Banks Edin	[Windermere: J. Garnett?]	[1854/59]	
F075	Rydal Water / Drawn & Engraved by W Banks Edin	[Windermere: J. Garnett?]	[1854/59]	
F079	Rydal Mount / Drawn & Engraved by W Banks and Son Edin	[Windermere: J. Garnett?]	[1854/59]	
F083	Old Mill Ambleside [Listed in contents as: Bridge at Ambleside] / Drawn & Engraved by W Banks and Son Edin	[Windermere: J. Garnett?]	[1854/59]	
F089	Brother's Water / Drawn & Engraved by W Banks and Son Edin	[Windermere: J. Garnett?]	[1854/59]	
F093	Upper Reach of Ullswater / L. Aspland delt; W. Banks sc, Edin.	[Windermere: J. Garnett?],	[1854/59]	

(Continued)

Table 5.1: Table of prints included in the Armytage and Follett manuscripts
(Continued)

Follett	Title / Artist, engraver	Publisher	Date	Series number
F097	Ara Force / L. Aspland delt; W. Banks sc, Edin.	[Windermere: J. Garnett?],	[1854/59]	
F105	Head of Derwentwater / W H Lizars	Kendal: John Hudson	[1854/59]	
F115	Vale of Keswick / L. Aspland delt; W. Banks sc, Edin.	[Windermere: J. Garnett?],	[1854/59]	
F119	Derwent Water / L. Aspland delt; W. Banks sc, Edin.	[Windermere: J. Garnett?],	[1854/59]	
F123	Waterfall at Lowdore / Drawn & Engraved by W Banks and Son Edin	[Windermere: J. Garnett?]	[1854/59]	
F131	Leathes Water / Drawn & Engraved by W Banks and Son Edin	[Windermere: J. Garnett?]	[1854/59]	
F133	Borrowdale / L. Aspland delt; W. Banks sc, Edin.	[Windermere: J. Garnett?],	[1854/59]	
F139	Wastwater / L. Aspland delt; W. Banks sc, Edin.	[Windermere: J. Garnett?],	[1854/59]	
F145	Ennerdale / Drawn & Engraved by W Banks and Son Edin	[Windermere: J. Garnett?]	[1854/59]	
F149	Crummock Water and Buttermere / Willm Banks; WH Lizars	Kendal: John Hudson	[1854/59]	
F153	Scale Force / Drawn & Engraved by W Banks and Son Edin	[Windermere: J. Garnett?]	[1854/59]	
F155	Honister Crag / L. Aspland delt; W. Banks sc, Edin.	[Windermere: J. Garnett?]	[1854/59]	
F165	Furness Abbey South East / Drawn & Engraved by W. Banks and Son Edin	[Windermere: J. Garnett?]	[1854/59]	
F173	Bassenthwaite Lake drawn & engd by W. B anks & Son, Edin.	[Windermere: J. Garnett?]	[1854/59]	

6 *The Marvels of Rome* (1889): The First English Translation of a Twelfth-Century Latin Description of the City Published by Ellis and Elvey of London

The *Mirabilia urbis Romae* is a much-copied, medieval, Latin text that served generations of pilgrims and tourists as a description of the city of Rome. The original manuscript dates from around 1140–3 and was written, it is thought, by a canon of St Peter's basilica by the name of Benedict.[1] It survives in numerous manuscript copies and is not to be confused with the similarly titled *Narracio de mirabilibus urbis Romae* of slightly later date (*c.* 1200) but whose author is known as the Anglo-Norman writer Master Gregory.[2] Unlike the *Mirabilia*, the *Narracio* remained undiscovered until 1917 and is known only in a single manuscript.

The *Mirabilia*, with which this chapter is concerned, was the most influential and popular medieval description of Rome from the mid-twelfth until the mid-fifteenth century. It has come down to us in a plethora of manuscripts found throughout medieval Europe, indicating

1 William Kynan-Wilson, 'Subverting the message: Master Gregory's reception of and response to the *Mirabilia Urbis Romae*', *Journal of Medieval History* 44:3 (2018), 347–64. The attribution is not universally accepted; for scholarship on this, the *Mirabilia's* relative reliability, and consideration to which literary genre it belongs, see Dale Kinney, 'Fact and Fiction in the *Mirabilia Urbis Romae*', in Éamonn Ó. Carragáin and Carol Neuman de Vegvar (eds), *Roma Felix – Formation and Reflections of Medieval Rome* (Aldershot: Ashgate, 2007), 235–52.

2 Master Gregorius, *Narracio de mirabilibus Urbis Romae*, late thirteenth-century copy on vellum, St Catherine's College, Cambridge, L V 87; M. R. James, 'Magister Gregorius de mirabilibus Urbis Romae', *The English Historical Review* 32:128, 1917, 531–54, here 531; Dale Kinney, Review of *Magister Gregorius, The Marvels of Rome*, trans. John Osborne, *Speculum* 64:4 (1989), 959–61.

both its rapid dissemination and extensive readership.[3] It is often cited as evidence of the twelfth-century renascence of interest in classical culture.[4] Its author, however, gave free rein to imaginative license, both in describing the monuments and locating amongst them tales of the early Christian martyrs. With the arrival of humanist learning, the contents of the text were consequently challenged. Particular traits of Roman humanism were its emphasis on archaeological and epigraphical studies, as taken up by Pomponio Leto (1428–98) and his academy, following examples (to name just two) set by Leon Battista Alberti (1404–72) in his *Descriptio urbis Romae* (c.1433) and Flavio Biondo (1392–1463) in his *Roma instaurata* (1444–9). Epigraphy had by now become a systematic and scientific subject of study, inscriptions appreciated for their value as ancient documents, therefore these scholars were keen to circulate their own accurate philological studies of Rome.[5] At the mercy of the *studia humanitatis*, the *Mirabilia* was thus eclipsed. The *Mirabilia* fell into relative obscurity, that is, until Francis Morgan Nichols (1826–1915) provided the first English translation, which was published in 1889 by Ellis and Elvey of London.[6]

This chapter aims, in Part One, to demonstrate the character of the *Mirabilia urbis Romae*, exemplifying some of its archaeological inaccuracies and hagiographical flights of fancy. It seeks, in Part Two, to understand the choice of this text for translation and publication, bearing in mind its discreditation by intervening academic scholarship, as outlined above. It posits that the selection by the translator and editor, Francis

3   Nina Robijntje Miedema, *'Die Mirabilia Romae': Untersuchungen zu ihrer Überlieferung, mit Edition der deutschen und niederländischen Texte* (Tübingen: Niemeyer, 1996), 24–95.
4   Erwin Panofsky, *Renaissance and Renascences in Western Art* (New York: Harper and Row, 1972), 73, 210.
5   Claudia La Malfa, 'Pinturicchio's Roman Fresco Cycles Re-examined 1478–1494' (PhD Thesis, Warburg Institute, University of London, 2003), 220.
6   Francis Morgan Nichols (ed. and trans.), *The Marvels of Rome or a Picture of the Golden City: An English Version of the Medieval Guide-Book with a Supplement of Illustrative Matter and Notes by Francis Morgan Nichols* (London: Ellis and Elvey; Rome: Spithöver, 1889).

Morgan Nichols, of this quasi-religious, medieval text was made for a blend of reasons, including the subject of his personal academic specialism, the business ethos of the Nichols family firm, and the more general nineteenth-century thrust to publish for posterity early historical material, whether from manuscript or early printed form. It takes into consideration, furthermore, the tastes and possible religious positionality of the publishers, Ellis and Elvey, who mixed with Arts and Crafts, Pre-Raphaelite and Symbolist painters and *literati*, and – self-evidently – with book collectors, some with Tractarian inclinations.

The methodology employed in the research has included textual analysis of the central primary source and comparative analysis in relation to other, later topographical renderings of Rome from the sixteenth century. Other primary sources consulted include John Ruskin's *Stray Letters from Professor Ruskin to a London Bibliopole*, a letter from Francis Morgan Nichols to the publishers, Ellis and Elvey, and copy correspondence from the publishers to their clients and to the printers, Nichols and Sons.[7] The cultural ambience of both medieval Rome and Victorian Britain are established through contextual analysis of other primary and secondary sources.

Part One: Rome and the *Mirabilia urbis Romae*

In the period from the middle of the fourth century to the end of the sixth, there had taken place in Rome a deliberate ecclesiastical policy of de-secularisation of church buildings and a promulgation of the cult of the saints. At the beginning of that period, Roman churches had borne

7   John Ruskin, *Stray Letters from Professor Ruskin to a London Bibliopole* [*sic*], ed. T. J. Wise, (London: privately printed, 1892). For the letter from Francis Morgan Nichols to the publishers, Ellis and Elvey, see University of California, Los Angeles (hereafter UCLA), UCLA Library Special Collections, Charles E. Young Research Library, Ellis Booksellers Records, Collection 425 Bx6 f6. For copy correspondence from the publishers to their clients and to the printers, Nichols and Sons, see UCLA Coll 425 Bx22, Bx23, Bx50.

the names of the donors who had facilitated their construction, such as the *titulus-Sabinae* on the Aventine Hill. In many cases, the donor's name became beatified, so that the *titulus-Sabinae* became the church of Santa Sabina.[8] During the fifth and sixth centuries the custom of dedicating churches to saints became the norm, with the Roman martyr saints – Peter, Paul, Laurence and Sebastian – being the most common. Hagiographical narratives were compiled by Roman clerics, fictionalizing the saints' lives and martyrdoms as having taken place in and around these churches.[9]

Alongside this went a trade in holy relics. Such was the belief in the capacity of relics to bring about miracles that the temptation of profiteering from a few old bones proved difficult for unscrupulous traffickers to resist. Even though strictly forbidden in Roman law, monks had been caught excavating pagan cemeteries, the bone fragments to be sold on as sacred relics.[10] The fact that relics were placed and venerated inside churches, many *within* the *pomerium* – the religious boundary of the city – brought about a revolution in early Christian funerary customs. The catacombs (which were cemeteries, and not underground places of worship for persecuted Christians) were located just outside of the city limits, since the dead could not be buried within the *pomerium*.[11] Increasingly, Christians wished to be buried in close proximity to the tombs of saints, martyrs and their relics, therefore burial within the walls of churches began to take place.[12] By Canon Benedict's time in the twelfth century, Rome's remaining classical ruins were construed by the Church as proof that the city had left behind its pagan ways.[13] The effect of this de-secularisation policy – not an

8   Bertrand Lançon, *Rome in Late Antiquity: Everyday Life and Urban Change, AD 313–604*, trans. Antonia Nevill, intro. Mark Humphries (Edinburgh: Edinburgh University Press, 2000), 159.

9   Lançon, *Rome*,159.

10 Lançon, *Rome*, 160.

11 Christopher Hibbert, *Rome: The Biography of a City* (London: The Folio Society, 1997), 64.

12 Lançon, *Rome*, 126–7.

13 Dale Kinney, 'Rome in the Twelfth Century: *Urbs fracta and renovatio*', *Gesta* 45:2 (2006) 199–220, here 215–6.

accidental one – was to attract a growing number of pilgrims to Rome. The city could draw on the double kudos of being both the *sedes Petri* and the architectural theatre in which the lives and deaths of the early Christian martyrs had been played out. This helped propagate the Seven Churches pilgrimage, the enclosed processional circuit to the indulgence-granting, 'Station' churches of the Jubilee pilgrimage, established, according to popular belief, by Pope Gregory I (r.590–604); the route of around sixteen miles was to be followed in a single day.[14]

Glancing ahead to the sixteenth century, this practice was to be revitalized by Filippo Neri (1515–95), founder of the Congregation of the Oratory, himself to be canonized after instigating reforms based on the Church's early Christian roots.[15] A commentator writing in 1559 recorded that:

> It is a joy to see the entire city of Rome, with such a cosmopolitan population, reformed with new life; its inhabitants filled with contrition, and all the churches filled with acts of devotion. On the roads leading to the Station churches [...], one no longer encounters people playing, joking, and ogling the ladies as they used to do. Instead, people now walk with lowered eyes, rosaries and books of hours in hand, saying their prayers.[16]

At the time in which the *Mirabilia* was written (*c.*1140–3), the inhabited part of Rome, the *abitato*, was a small medieval city located in the bend of the River Tiber. Between the *abitato* and the sprawling, encircling ruins of the walls and gates of the great ancient city now lay vineyards, marshy waste ground and pastureland – the *disabitato* – in which animals grazed among the temples and arches, giving to the Roman forum its nickname

---

14 Minou Schraven, 'Roma Theatrum Mundi: Festivals and Processions in the Ritual City', in Pamela M. Jones, Barbara Wisch and Simon Ditchfield (eds), *A Companion to Early Modern Rome, 1492–1692* (Leiden: Brill, 2019), 247–65, here 259.

15 Schraven, '*Roma Theatrum Mundi*', 254.

16 *Avvisi delle cose nuove successe in Rome et del governo della Città, et buon ordine di vivere, et dei sepolchri, processioni, orationi, et altre opere pie fatte dalle Confraternitie et Compagnie di Roma et altre cose nuove* (Rome, 1559), unpaginated; cited by Schraven, '*Roma Theatrum Mundi*', 255.

of *campo vaccino* (the cow pasture).[17] In the preceding centuries, as summarized above, Rome had been transformed from the capital of the Roman Empire to the ecclesiastical nerve centre of the Christian west. The city's topography reflected that change, churches now superseding in importance the monuments and edifices of the classical past. Church buildings, the papacy, saints' cults and visiting pilgrims constituted the palimpsest city's overlaid identity.[18] From the *Mirabilia*, however, emerges a shadowy image of the ruins remaining in Rome in the twelfth century, many of which have since disappeared; in fact, the book contributed to the preservation of some of the constructions that may otherwise have been lost.[19]

The earliest extant manuscript copy of the *Mirabilia* is dated to the end of the twelfth century.[20] Nichols' English translation *The Marvels of Rome or a Picture of the Golden City* published in 1889 follows the work in its original form by falling into three parts: Part I, the foundation of Rome and a summary of sites arranged under various headings (gates, arches, hills, etc.), Part II, a collection of legends associated with extant, ancient monuments, and Part III, a walk around the city, beginning at the Vatican and ending in Trastevere. For the purpose of this paper, sample passages will be examined from each part, using Francis Nichols' 1889 translation for quotations.

Part I of the book, entitled 'Of the Foundation of Rome; and of her Wall, Gates, Arches, Hills, *Thermae*, Palaces, Theatres, Bridges, Pillars, Cemeteries, and Holy Places', opens by tracing the origins of Rome to the descendants of Noah. Citing Marcus Terentius Varro (116–27 BC), considered ancient Rome's most learned polymath, our author names the individuals who established settlements on the seven hills: biblical, historical or mythological figures including Hercules, and Roma, daughter of

---

17 Hibbert, *Rome: The Biography*, 155. 'The layout of a good many western towns fluctuated inside their ancient pomerium as if in clothes that were too big for them'; Lançon, *Rome*, 15.

18 Mark Humphries, 'Introduction: The city between antiquity and the Middle Ages', in Lançon, Rome, xvi–xxii.

19 Nichols (ed. and trans.), 'Preface', *Marvels of Rome*, v–xxvi, here vi.

20 Biblioteca Apostolica Vaticana (hereafter B.A.V.), Vat. Lat. 3973, fols 1r–14v.

Aeneas, among many others. It relates the birth of Romulus, King of the Trojans, who is credited with encircling the seven hills with a wall, naming the aggregation Rome, after himself.[21]

Canon Benedict seems to have revelled in statistics. The first feature described, 'Of the Town Wall', '[...] hath towers three hundred threescore and one, castles forty and nine, [chief arches seven,] battlements six thousand and nine hundred, gates twelve, posterns five; and in the compass thereof there are twenty two miles', without counting in Trastevere and the 'Leonine city' (the Vatican).[22] Similarly, under 'Of the Pillars of Antonine and of Trajan', the reader is informed that:

> In Rome were twenty and two great horses of gilded brass, horses of gold fourscore, horses of ivory fourscore and four, common jakes [public privies] an hundred and fourscore, great sewers fifty, bulls, griffons, peacocks, and a multitude of other images, the costliness of whereof seemed beyond measure, insomuch that men coming to the city had good cause to marvel at her beauty.[23]

Benedict was relying for his numerical exactitude on a military and administrative document of about the end of the fourth century; evidently he was impressed by the former splendour of ancient Rome and wished to impart his awestruck astonishment to his reader.[24]

The text is equally well-endowed with anecdote. In a section in Part I devoted to 'Of the Gates', the *portae* are named and located: '*porta Appia*, [where is the church, that is named *Domine quo Vadis*, that is to say, Lord, whither goest thou, where are seen the footsteps of Jesus Christ]'.[25] According to legend, in AD 56, the apostle Peter, fleeing persecution in Rome and travelling south along the Via Appia, is said to have met his

---

21 Nichols (ed. and trans.), *Marvels of Rome*, 1–5. In all cases, the square brackets are Nichols' edits.
22 Nichols (ed. and trans.), *Marvels of Rome*, 6. The Leonine walls of the Vatican were built by Leo IV (790–855).
23 Nichols (ed. and trans.), *Marvels of Rome*, 26.
24 J. K. Hyde, 'Medieval Descriptions of Cities', *Bulletin of the John Rylands Library*, 38 (1965–6), 308–40, here 321.
25 Nichols (ed. and trans.), *Marvels of Rome*, 7.

Saviour moving in the opposite direction and asked him, 'Where are you going?' Jesus answered that he was going to Rome to be crucified anew. The Church of Domine Quo Vadis was built at the site of their supposed meeting. Peter was captured, returned to Rome and put to death. The aforementioned footprints of Christ are in fact a reproduction; the originals are purported to be found in the pilgrimage church of San Sebastiano on the Via Appia.[26] The *Domine Quo Vadis* is mentioned again under 'Of places where Saints suffered', the final section of Part I; 'These are the places that are found in the passions of Saints: without the Appian gate, the place where the blessed Sixtus was beheaded, and the place where the Lord appeared to Peter, when he said, Lord, whither goest thou.'[27] Thus Canon Benedict wove narrative detail from the lives of the early Christian saints around the ancient classical buildings that were still extant in the twelfth century.

A further splash of colour is brought to this otherwise generally monochrome list of locations on mention of the following detail; 'the *acqua Salvia* at Sant'Anastasius, where the blessed Paul was beheaded, [and the head thrice uttered the word Jesus, as it bounded, and where there be yet three wells which spring up diverse in taste]'.[28] The sudden, miraculous appearance of springs was an accepted signifier of Christian sanctity – in this case visual, terrestrial proof of the holy martyrdom of Saint Paul. A different kind of waterway marked the location of St Sebastian's demise: 'the hill of Scaurus, which is between the Amphitheatre [Colosseum] and the Racecourse [Circus Maximus], before the Seven Floors [Septizonium],[29]

---

26 Eileen Gardiner, 'Gazetteer', in Francis Morgan Nichols (ed. and trans.), *The Marvels of Rome: Mirabilia urbis Romae*, 2nd edn (New York: Italica Press, 1986), 51–111, here 67–8.

27 Nichols (ed. and trans.), *Marvels of Rome*, 29.

28 Nichols (ed. and trans.), *Marvels of Rome*, 30.

29 The Septizonium, demolished by Pope Sixtus V (1585–90), resembled a colossal *scenae frons* of a Roman theatre, built against the south-eastern slope of the Palatine so that it could be seen by travellers approaching Rome from the south; Hibbert, *Rome: The Biography*, 340.

where is the sewer, wherein Saint Sebastian was cast.'[30] Sebastian (third century), probably a military officer, once his beliefs became known to the authorities, was shot by archers so that, as the thirteenth-century hagiographic account in *The Golden Legend* tells us, 'the saint was surrounded by arrows as a porcupine is with quills.'[31] While offering a pretext in Renaissance art for the depiction of the idealized male nude, this saint's body in medieval art is sometimes barely discernible beneath the bristles of profuse arrows, notwithstanding the fact that these were not ultimately fatal.[32] Continuing the tale, *The Golden Legend* reports that, later, returning to the emperor to re-state his faith, Sebastian was ordered to be cudgelled to death and his body thrown into the Cloaca Maxima, Rome's main sewer, as recounted in the *Mirabilia*.[33] Believing that disease was caused by the arrows of Apollo, the ancients found in the punctured body of Saint Sebastian an appropriately formed intercessor through whom to pray for protection from the plague. Thus began his cult in the fourth century, when the basilica of San Sebastiano was constructed above his tomb in the catacombs on the Via Appia.[34]

While Benedict's statistical information, such as the number of gates and statues, derived from monotonous catalogues of extant monuments, the more engaging flashes of colour pertaining to places of martyrdom of the Roman saints emanate from Christian tourist literature. Both kinds of source material were being written and repeated from the fourth century onwards, to satisfy the voracious appetite of curious visitors.[35]

30 Nichols (ed. and trans.), *Marvels of Rome*, 31.
31 Jacobus de Voragine, *The Golden Legend: Readings on the Saints*, trans. William Granger Ryan, 2 vols (Princeton NJ: Princeton University Press, 1993), Vol. 1, 97–101, here 97.
32 Such as Giovanni del Biondo's *Martyrdom of Saint Sebastian* (*c.*1370), Florence: Museo dell'Opera del Duomo.
33 Sebastian was thrown into the sewer, 'to prevent the Christians from honouring him as a martyr'; de Voragine, *The Golden Legend*, 100.
34 James Hall, *Dictionary of Subjects and Symbols in Art* (London: John Murray, 1974), 276–7.
35 Hyde, 'Medieval Descriptions', 320–1.

Part II of the *Mirabilia*, as translated by Nichols, is subtitled 'The Second Part containeth divers Historics touching certain famous Places and Images in Rome', or, in today's parlance, the legends behind Rome's monuments.[36] The second section of this part, 'Of the Marble Horses [the *Dioscuri* on the Quirinal]', is here taken as a case study. This comprises two paragraphs; an introduction to the sculptures, followed by the narrative which purportedly accounted for their form (Figure 6.1). A passage from the second paragraph, the longest, reads thus:

> In the time of emperor Tiberius there came to Rome two young men who were philosophers, named Praxiteles and Phidias, whom the emperor, observing them to be of so much wisdom, kept nigh unto himself in his palace; and he said to them, wherefore do ye go abroad naked? who answered and said: Because all things are naked and open to us, and we hold the world of no account, therefore we go naked and possess nothing; and they said: Whatsoever thou, most mighty emperor, shalt devise in thy chamber by day or night, albeit we be absent, we will tell it thee every word. If ye shall do that ye say, said the emperor, I will give you what thing soever ye shall desire. They answered and said, We ask no money, but only a memorial of us. And when the next day was come, they showed unto the emperor in order whatsoever he had thought of in that night. Therefore he made them the memorial that he had promised, to wit, the naked horses, which trample on the earth, that is upon the mighty princes of the world that rule over the men of this world; and there shall come a full mighty king, which shall mount the horses, that is, upon the might of the princes of this world. Meanwhile there be the two men half naked, which stand by the horses, and with arms raised on high and bent fingers tell the things that are to be; and as they be naked, so is all worldly knowledge naked and open to their minds.[37]

36 Nichols (ed. and trans.), *Marvels of Rome*, 35.
37 Nichols (ed. and trans.), *Marvels of Rome*, 39–41. The philosophers claimed that the princes of the world were like horses which had not yet been mounted by a true king; Francis Haskell and Nicholas Penny, *Taste and the Antique: The Lure of Classical Sculpture 1500–1900* (New Haven and London: Yale University Press), 136–41, here 136.

Figure 6.1: Anon, *Statues of The Dioscuri at the Quirinal, Rome*, from Antoni Lafréry (*1512–77*), *Speculum Romanae Magnificentiae*, (Rome: Claude Duchet, 1584). Engraving. New York; Met Museum, Accession Number 17.50.19–122.

In the public domain.

Identified by subsequent antiquarian scholarship as the demi-gods Castor and Pollux, twin sons of Leda, these sculptural figures were among the few colossal statues visible in Rome throughout the Middle Ages. Then known as *The Horse-Tamers of Monte Cavallo*, the group had endowed the Quirinal with its local equestrian name.[38] Eighteen feet tall, these two muscular, nude giants strain to control their rearing steeds, their arms raised and fists clenched. These, as we now understand, denote tautly held reins, although Canon Benedict in his *Mirabilia* (as in the translation above) construed these as 'arms raised on high and bent fingers [that] tell the

38 Phyllis Pray Bober and Ruth Rubinstein, *Renaissance Artists and Antique Sculpture: A Handbook of Sources* (London: Harvey Miller, 1986), 159–61.

things that are to be'.[39] The sculpted figures' heads sharply turn and their bodies twist energetically as they struggle to subdue the recalcitrant animals, whose veins throb and nostrils flare with their brute force. The view of the group in Antoni Lafréry's *Speculum Romanae Magnificentiae* of 1584 shows the power communicated by the figures' anatomies. The damaged forequarters of the horse on the right were, however, at the time of this engraving, supported on blocks made of brick.

On separate bases but united at the time of the engraving by a small dwelling in-between, the inscriptions, thought to have been carved in the fourth century AD, are – left – 'OPUS PRAXITELIS', and – right – 'OPUS FIDIAE', the works attributed respectively to the two Greek sculptors, Praxiteles and Phidias.[40] Made famous for posterity by Pliny the Elder (d. AD 79), who wrote the earliest extant account of Greek art in an appendix to his *Natural History*, the names were apparently mistaken by the author of the *Mirabilia* to denote the *subject matter* of the sculptures (in his reading, the two philosophers), rather than the names of the sculptors. According to Canon Benedict, the philosophers were so accurate in their psychic readings of the emperor that he rewarded them with the building of a monument of their choice. The humanist Flavio Biondo scoured a vast range of Latin texts to try to trace the sculptures' provenance, followed up by antiquarians Pirro Ligorio and Onofrio Panvinio in the sixteenth century, though it was Alessandro Donati in a late edition of his *Roma vetus ac recens* (1665) who identified the figures as the *Dioscuri*, Castor and Pollux, which was finally accepted in the nineteenth century.[41] Part II of the *Mirabilia* is thus shown to contain erroneous information, though as our translator, Francis Nichols, forgivingly points out in his footnote, 'The legend of Phidias and Praxiteles, and that which follows in

39   Nichols (ed. and trans.), *Marvels of Rome*, 39 n93.
40   Bober and Rubinstein, *Renaissance Artists*, 159. Between 1589 and 1591, under Sixtus V, they were restored and given new bases with a fountain between them. In 1786, under Pius VI, the ancient obelisk that we see today was placed between them; Haskell and Penny, *Taste and the Antique*, 136.
41   Bober and Rubinstein, *Renaissance Artists*, 159–60; Haskell and Penny, *Taste and the Antique*, 139.

the next chapter [...] are evidently stories which had their origin upon the spot, out of the fancy of pilgrims, or of their guides.'[42] This is one example of many misconceptions contained in this section of the *Mirabilia*, thus without the benefit of Nichols' footnotes, the original text is seen to have been misleading in parts regarding the remains of twelfth-century Rome.

Part III of the book 'containeth a Perambulation of the City'.[43] Constituting two-thirds of the *Mirabilia*, this part is described by J. K. Hyde as 'speculative archaeology', which he considered represents Benedict's original ideas.[44] While in Part II, Benedict had aimed to establish a link between the pagan past and the Christian era, here in Part III reference to religious figures is negligible. In a section entitled 'Of the parts of the City nigh unto the Tiber', however, the author writes:

> At Saint Stephen *in Piscina* (that is to say, at the Cistern) the palace of the Prefect Chromatius, and a temple that was called *Holovitreum*, being made out of glass and gold by mathematical craft, where was an astronomy [astrograph] with all the signs of the heavens, the which was destroyed by Saint Sebastian with Tiburtius, the son of Chromatius.[45]

To amplify the narrative, the prefect Chromatius was suffering from gout but was promised recovery if he converted to Christianity and gave permission to St Sebastian to destroy all of the pagan artefacts in his house. Chromatius's father had constructed a gold and crystal chamber – the '*cubiculum holovitreum*' – which contained automated images of the zodiac, enabling Chromatius to predict the future. He was reluctant to allow its destruction, however, did so and was restored to full health.[46] The narrative comes from the late antique *Acts of St Sebastian*; Benedict is thus again drawing on Christian tourist literature for the story.

42 Nichols (ed. and trans.), *Marvels of Rome*, 39 n93.
43 Nichols (ed. and trans.), *Marvels of Rome*, 70–117.
44 Hyde, 'Medieval Descriptions', 322. This view is endorsed by Kinney, 'Fact and Fiction', 236.
45 Nichols (ed. and trans.), *Marvels of Rome*, 114.
46 Kinney, 'Rome in the Twelfth Century', 217, citing Acta S. Sebastiani, XVI, 57, col.1046.

Part One of this chapter has shown that Canon Benedict's *Mirabilia urbis Romae* catered for the appetites of late medieval and early modern pilgrims and tourists, while simultaneously providing later writers with material for more practicable guidebooks and further, more forensically researched descriptions of Roman topography.[47] The above sample excerpts demonstrate that Benedict was drawing on surviving architectural remains, classical texts such as Varro, Ovid and Livy, military and administrative records, Christian tourist literature, his own fertile imagination and his enthusiasm for the achievements of the ancients. This chapter's Part Two will consider the identity and family background of Francis Morgan Nichols, the editor and translator of the *Mirabilia*, his relationship with the publisher, Ellis and Elvey and possible reasons why this text was selected for translation and publication.

Part Two: Francis Morgan Nichols, Ellis and Elvey and *The Marvels of Rome*

The great Italian historian Ludovico Muratori (1672–1750) had recognized the significance of the literary genre of descriptions of cities and expressed a desire that more of these should be made available to historians in printed form.[48] The translator and editor of the *Mirabilia urbis Romae*, published in 1889 as *The Marvels of Rome or a Picture of the Golden City*, was Francis Morgan Nichols (1826–1915), a member of the Nichols dynasty of printers. For several generations the family had been central to topographical and

---

47 Hyde, 'Medieval Descriptions', 323–38. The shop of Antonio Blado (1490–1567) was nonetheless still printing the Mirabilia under the title of *Le cose maravigliose dell'alma città di Roma* through the 1560s; Evelyn Lincoln, 'Printers and Publishers in Early Modern Rome', in Pamela M. Jones, Barbara Wisch, Simon Ditchfield (eds), *Companion to Early Modern Rome 1492–1692* (Leiden: Brill, 2019) 546–63, here 549.

48 Hyde, 'Medieval Descriptions', 308, citing Ludovico Muratori, *Rerum Italicarum Scriptores* (Typographia Societatis Palatinae: Milan, 1727), 11:3.

antiquarian scholarship, able to publish and print their own research.[49] John Nichols (1745–1826) was the first in the family line to run this, one of London's largest printing houses; its output and reputation made it the preferred choice of antiquaries across the land, especially given Nichols's editorship and publication of the most distinguished literary periodical in Britain, *The Gentleman's Magazine*.[50] He was succeeded in turn by John Bowyer Nichols (1779–1863) and John Gough Nichols (1806–73), the latter Francis Morgan's eldest brother.[51] John Gough Nichols was a founding member in 1834 of the Durham-based Surtees Society and, borne along by its success, he helped inaugurate in 1838 the Camden Society, whose motivations included the encouragement of local historical research and ensuring the longevity of documentary evidence, hence the publication of manuscripts and new editions of rare printed books.[52] These commitments likewise spurred John on to participate in the foundation of the Royal Archaeological Institute (1844), the London and Middlesex Archaeological Society (1855), and other such bodies. The family press under John Gough Nichols printed for all of these societies.[53] The company also had printing contracts with, amongst others, Parliament, the Society of Antiquaries, and the Royal Society. By the third generation of the firm, John Gough Nichols, Francis's brother, was at the epicentre of a web of

49 The author is indebted to Barry McKay of Barry McKay's Rare Books for identifying this family relationship. On three generations of the Nichols family, see Julian Pooley, 'The Nichols Family and their Press (1777–1873)', <https://www.bl.uk/picturing-places/articles/the-nichols-family-and-their-press> accessed 3 July 2022.

50 The family produced the *Gentleman's Magazine* from 1778 to 1856; Julian Pooley, 'The Book that Changed my Life: Discovering the Archive behind the Gentleman's Magazine', *The Book Collector* (Spring 2021), 2–16.

51 Julian Pooley, 'Beyond the Literary Anecdotes: The Nichols Family Archive as a Source for Book Trade Biography', in Robin Myers, Michael Harris and Giles Mandelbrote (eds), *Lives in Print: Biography and the Book Trade from the Middle Ages to the 21st Century* (London: British Library, 2002), 119–50. The family tree is Figure 1, 120.

52 On the Surtees and Camden Societies, see Philippa Levine, *The Amateur and the Professional: Antiquarians, Historians and Archaeologists in Victorian England, 1838–86* (Cambridge: Cambridge University Press, 1986), passim.

53 Pooley, 'The Nichols Family'.

local and national institutions given over to topography, archaeology, literary biography, local history and genealogy, hence a national network of erudite scholars.[54]

Francis Morgan Nichols had graduated from Wadham College, Oxford, becoming a Barrister at Law at Lincolns Inn Fields, and went on to produce a modest publication record in law.[55] Following in the footsteps of his forebears and their circle, however, his abiding interests appear to have been local history and the topography and archaeology of ancient Rome.[56] Like other members of his family, he conducted his own research and gave presentations to the Society of Antiquaries, which were then published by the Nichols press.[57] In Francis's later years, he translated, edited and commented on the letters of the Catholic theologian Desiderius Erasmus of Rotterdam, the first two volumes appearing during his lifetime, the third posthumously.[58]

Francis's aptitude for languages was already apparent early in his career; a letter to his brother, John Gough Nichols, dated 12 March 1851 (when Francis was twenty-five), records that he was busy working on a

54 Julian Pooley, 'Printing the Past: Discovering the Archive of the Nichols Family of Printers, Antiquaries and Editors of the Gentleman's Magazine, 1777–1873', Franco-British History Seminar of the Sorbonne, Paris and Institute of Historical Research Séminaire franco-britannique d'histoire, 6 May 2021 <https://www.youtube.com/watch?v=vuCWTjyirgc> accessed 6 July 1922.

55 For example, John le Breton (d.1275), *Britton [on the Laws of England]*, trans. from French, ed. and intro. by Francis Morgan Nichols, 2 vols (Oxford: Clarendon Press, 1865).

56 For example, Francis Morgan Nichols, *The Hall of Lawford Hall: Records of an Essex House and of its Proprietors from the Saxon Times to the Reign of Henry VIII* (London: Ellis and Elvey, 1891); Francis Morgan Nichols, *The Roman Forum: A Topographical Study* (London: Longmans and Co., 1877).

57 For example, Francis Morgan Nichols, *The Regia, the Atrium Vestae and the Fasti Capitolini, Communicated to the Society of Antiquaries* (Westminster: Nichols and Sons, 1887).

58 Desiderius Erasmus, *The Epistles of Erasmus: from his Earliest Letters to his Fifty-First Year, Francis Morgan Nichols* (trans. and ed.) 3 vols, Vol. I (London: Longmans, Green and Co., 1901); Vol. II (London: Longmans, Green and Co., 1904); Vol. III (London: Longmans, Green and Co., 1918).

translation, possibly John le Breton's text of 1275 on the laws of England that was published in 1865.[59] With a facility, then, in the romance languages, Francis translated the *Mirabilia urbis Romae* from Latin into English; *The Marvels of Rome* was published – not by the Nichols press – but by Ellis and Elvey of London and Spithöver of Rome. The book, like many of his publications cited above, was placed with a reputable publisher and then directed to his brother's firm, J. B. Nichols and Sons at 25, Parliament Square, Westminster, for printing. Nichols and Sons did on a number of occasions publish works, but only when they felt comfortable with the financial risk; they preferred to liaise with authors on matters such as copy, paper, print run and illustrations. The Nicholses thus regarded themselves printers rather than publishers, leaving the latter responsible for distribution and sales.[60] The additional involvement here of Spithöver is logical in view of the book's Roman subject matter; Francis already had an established connection with this publisher.[61] The Nichols archive is moot regarding Francis's selection of Ellis and Elvey as his British publisher, however there was a family relationship with the firm through the person of Robert Victor Elvey (1858–1934). He was Francis's nephew, the son of Isabella Georgiana Nichols (1821–63) and Sir George Job Elvey (1816–93), the organist at Windsor Castle.[62] Robert Elvey was taken into partnership by Gilbert Ifold Ellis in 1887; further detail concerning Ellis and Elvey follows below. Francis may thus have channelled publication of *The Marvels of Rome* in that direction to benefit his nephew's nascent partnership.[63]

As established in Part One of this chapter, the *Mirabilia urbis Romae* is a part-fictionalized and imprecise description of Rome, compiled in a period prior to the philological scrutiny that had come with the humanist

59 Folger Shakespeare Library, Washington (Y.d.24 12B/17 NAD13567); the author is indebted to Julian Pooley for this information.
60 Emails to author from Julian Pooley, 10 and 28 July 2022.
61 Francis Morgan Nichols, *Notizie dei rostra del Foro Romano e dei monumenti contigui* (Rome: Spithöver, 1885).
62 Smith and Benger, *The Oldest London Bookshop*, 70.
63 Email to author from Julian Pooley, 10 July 2022.

movement. This might give rise therefore to the question as to why this text was selected by Francis Nichols for his translation and publication. That which would blossom into Francis's fascination with classical, archaeological remains is already foreshadowed in a letter to his eldest sister, Mary Anne Iliffe Nichols (1813–70), written in 1851 during a sojourn in Italy. He was clearly intoxicated by the romantic beauty of Venice and continues to describe, amongst other things, an almost perfectly preserved Roman amphitheatre in Verona, significantly commenting on the fact that it was once used for gladiatorial performances and the 'cruel deaths of Christian martyrs'.[64] That he mentions this indicates at least his passing interest in the fate of the early saints. This seems to be confirmed by his choice of frontispiece for *The Marvels of Rome*, a close-up photograph of a panel from the bronze portal of St Peter's (Figure 6.2).

Figure 6.2: Frontispiece to Francis Morgan Nichols (ed. and trans.), *The Marvels of Rome or a Picture of the Golden City* (London: Ellis and Elvey; Rome: Spithöver, 1889). Photogravure of Antonio Filarete's *Portal to St Peter's* (detail), bronze, 1433–45.

64 Nichols Archive Database, Private Collection 1, PC1/54/45/1-2 NAD6620. The author is grateful to Julian Pooley for this information.

Francis describes the frontispiece at the very end of *The Marvels of Rome*:

> The bronze doors of St. Peter's, made for Eugenius IV. in 1447, have among other ornaments a bas-relief of the Passion of St. Peter by Antonio Filarete. In this work, to mark the locality, the foreground is occupied by a row of objects conceived in the spirit of the *Mirabilia*. These are the 'Sepulchre of Remus' with a figure of Roma before it, the Tiber with shields and arms floating on it, the 'Temple of Hadrian', the Terebinth, and the 'Sepulchre of Romulus'. The last three objects symbolize the place of Saint Peter's crucifixion.[65]

As in the whole book, details in this passage are amply annotated, which cross-refer the reader to salient points in the main text. Francis moreover shows himself to be thoroughly acquainted – not only with his immediate Latin sources – but with classical literature and the actual topography of Rome, exemplified in the following footnotes:

> This double name [Aventinus Silvius] is taken from Varro, supplemented by Livy. *Aventinum ...* (dictum) *a rege Aventino Albano.* Varro, L. L. 43. *Mansit Silviis postea omnibus cognomen qui Albae regnaverunt.* Liv. i.3.[66]

And elsewhere:

> The Porta Collina of the *Mirabilia* and *Ordo Romanus* is not this gate [Porta Castelli], but that closing the bridge on the side of the Borgo.[67]

As already witnessed, Francis published several texts on the classical ruins of ancient Rome. Extant pencil and ink sketches by his own hand, containing careful notes and instructions to an (unknown) engraver, as well as a few proofs, testify to Francis's study of surviving monuments in preparation for illustrating his academic papers.[68] Francis would certainly have read

---

65 Nichols (trans. and ed.), *The Marvels*, 196.
66 Nichols (trans. and ed.), *The Marvels*. 4, n.9.
67 Nichols (trans. and ed.), *The Marvels*, 196 n.5.
68 Nichols Archive Database, Private Collection 1, PC1/34/46 NAD5286; Private Collection1, PC1/34/47 NAD5287; Private Collection 1, PC1/34/50 NAD5290; Private Collection 1, PC1/34/51 NAD5291. Data provided by Julian Pooley.

the *Mirabilia* in the course of his own research. Despite its pre-humanist idiosyncrasies, it had remained a staple of the Roman printing houses, catering for Rome's ongoing pilgrimage and tourist trades.[69] The book would thus, firstly, have added fuel to Francis's antiquarian appetite, as well, secondly, as his curiosity about Rome's holy figures. Thirdly, its translation and publication would have met with the traditional Nichols firm's commitment to facilitate wider access to scarce material. Francis's grandfather, John Nichols, in Nichols-scholar Julian Pooley's words, had a 'zeal [...] to preserve in print the fugitive writings of authors shunned by the canon of literature',[70] while Francis's eldest brother, John Gough Nichols, had mobilized the same ethos in being a founder member of the learned societies mentioned above, whose local historical and topographical research was printed by his press.

It is germane to note that among these local institutions there was a strong relationship between memberships, theology and church architecture. In the year of the formation of the London Camden Society (1838), for example, thirty-six per cent of its members were already Fellows of the Society of Antiquaries and twenty per cent clergymen. Of the Surtees Society, the respective figures were ten per cent and twenty-five per cent.[71] It is not surprising, therefore, that John Gough Nichols found himself at the nexus between local historians and topographers on one hand, and clergy interested in the restoration of their churches on the other.[72] The surge of interest in England in the Gothic Revival style can be traced back to the Tractarian movement of the 1830s that came out of Oxford and spread like ripples in a pond. Members of this 'Oxford Movement', as it came to be known, argued for the need of the Church of England to rediscover and demonstrate its Catholic roots. The Cambridge Camden

69 Lincoln, 'Printers and Publishers in Early Modern Rome', 549–50, 557. Contemporary chronicles estimate that up to 50,000 pilgrims visited Rome for Easter week in a jubilee year, presenting an ongoing demand for such texts.
70 Pooley, 'Printing the Past' (at around 36 mins).
71 Levine, *The Amateur*, 44.
72 Pooley, 'The Nichols Family', under 'John Gough Nichols'.

Society founded in 1839 – in 1845 to become the Ecclesiological Society – promoted Victorian Gothic as the aesthetic and architectural ideal, as representing the truest expression of Roman Catholic-inclined Anglicanism and its 'high church' values: this at a time when antique culture and the high Renaissance style were still very much in vogue.[73] The Cambridge Camden Society translated and published the grand *opus* of the medieval canonist Guillaume Durand, his encyclopaedic *Rationale Divinorum officiorum*, re-titled in its English version *The Symbolism of Churches and Church Ornaments*; this was to form the standard source for neo-Gothic, ecclesiastical restoration and furnishing.[74] Clearly, the Gothic Revival style was not relevant to Francis Morgan Nichols' archaeological interests in classical Rome, however the sweeping tide of medievalism may have increased his attention to the twelfth-century text of the *Mirabilia urbis Romae*.[75] Ellis and Elvey – Francis's publishers of *The Marvels of Rome* – were themselves far from untouched by the Victorian preoccupation with the Middle Ages and this can already be seen at work in relation to the founder of the firm, Frederick Startridge Ellis (1830–1901). For this reason, it is helpful to now sketch in some basic background about the origins of Ellis and Elvey and the intellectual ambient of its clientele.

In 1860, self-educated Frederick Ellis opened an antiquarian bookshop in Covent Garden.[76] In 1872 he took over the property of T. and W. Boones

73   Michael Chandler, *An Introduction to the Oxford Movement* (London: SPCK, 2003), 109; Joanna Banham, 'William Morris, Medievalism, and the Chivalric Ideal' in Melissa E. Buron (ed.), *Truth and Beauty: The Pre-Raphaelites and the Old Masters* (Munich, London, NY: Fine Arts Museum of San Francisco and DelMonico Books, 2018), 209–16, here 209.

74   Martin Ellis, 'Practice and Ideal; Ideas and Objects, 1850–1910', in Martin Ellis, Victoria Osborne and Tim Barringer (eds), *Victorian Radicals: From the Pre-Raphaelites to the Arts and Crafts Movement* (New York: American Federation of Arts and DelMonico Books, 2018), 53–69, here 57–9.

75   Joanne Parker and Corinna Wagner (eds), *The Oxford Handbook of Victorian Medievalism* (Oxford: Oxford University Press, 2020).

76   At 33, King Street; Smith and Benger, *The Oldest London Bookshop*, 65; William Roberts, *The Book Hunter in London: Historical and Other Studies of Collectors and Collecting* (London: Elliott Stock. 1895), 245–6.

at 29, New Bond Street – the oldest bookshop in London – whose origins could be traced back to 1728 (Figure 6.3).[77] Purchasing also its goodwill, Ellis succeeded Boones as buyer for the British Museum, showing himself to be a shrewd and talented businessman with discriminating taste and geniality. Building up a clientele of influential *literati*, he introduced William Morris (1834–96) to the attractions of illuminated manuscripts, woodblock prints and incunabula,[78] which became so fundamental in his design and craftsmanship.[79] Ellis had already published Dante Gabriel Rossetti's first volume of original verse (1870).[80] The Romantic poet and novelist Algernon Swinburne, moving in Pre-Raphaelite circles, chose Ellis as his publisher.[81] This circle was united in its admiration of John Ruskin, in particular his chapter on 'The nature of Gothic', in *The Stones of Venice* (1851–3), which had drawn attention to the decline of post-medieval society.[82] These early business contacts led to lifelong friendships. Ellis was soon on amicable terms with aesthete and symbolist painter and designer, Edward Burne-Jones (1833–98), who – along with Morris – had studied medieval manuscripts at the Bodleian and the British Library.[83] It is germane to remember that these two eminent artist-designers had met at Oxford in the mid-1850s during the apogee of the Tractarian movement, both studying for a degree in theology, which they had both later abandoned. Morris and Burne-Jones remained friends, working collaboratively with

77 Smith and Benger, *The Oldest London Bookshop*, 2; Roberts, *The Book Hunter in London*, 245–6.
78 Smith and Benger, *The Oldest London Bookshop*, 65.
79 Michaela Braesel, 'The Influence of Medieval Illuminated Manuscripts on the Pre-Raphaelites and the Early Poetry of William Morris', *Journal of William Morris Studies* 15/4 (2004), 1–10.
80 Smith and Benger, *The Oldest London Bookshop*, 65–6. Dante Gabriel Rossetti, Poems (London: F. S. Ellis, 1870).
81 Smith and Benger, *The Oldest London Bookshop*, 65–6.
82 Jennifer Harris, 'William Morris and the Middle Ages', in Joanna Banham and Jennifer Harris (eds), *William Morris and the Middle Ages* (Manchester: Manchester University Press, 1984), 1–17, here 6.
83 His study of fourteenth- and fifteenth-century miniatures is recognisable in Burne-Jones' work; Braesel, 'The influence of Medieval Illuminated Manuscripts', 2.

Figure 6.3: The Interior of Ellis's Bookshop at 29, New Bond Street. Photograph, courtesy of George Smith and Frank Benger, *The Oldest London Bookshop: A History of Two Hundred Years*, (London: Ellis, 1928), frontispiece.

Rossetti, or independently, in a medieval style, populating their designs with chivalric and romantic figures.[84] Poet laureate and an 'immortal' of the Pre-Raphaelites, Alfred Lord Tennyson (1809–92) frequented Ellis's shop, as did James Abbott McNeill Whistler (1834–1903), the American-born artist who was to become the leading protagonist of the notion of 'art for art's sake'.[85] The majority of the most renowned collectors of the period

84 For example, the episodes from the legend of King Arthur on the murals of the Debating Chamber of the Oxford Union Society: later in decorating Morris's Red House; Ellis, 'Practice and Ideal', 59.
85 Smith and Benger, *The Oldest London Bookshop*, 65–6.

were Ellis's clients. These included bibliophiles such as John Cole Nicholl (1823–94)[86] and Sir William Stirling-Maxwell (1818–78).[87]

Frederick Ellis was knowledgeable on a diverse range of subjects, though his interests were most closely aligned with those of William Morris.[88] Between 1874 and 1884, the two were joint leaseholders of Kelmscott Manor, where Ellis happily spent his weekends in Morris's company. As poet, designer, craftsman and political thinker, Morris drew on the Middle Ages more consistently than any of his colleagues and peers.[89] Right up until Morris's death (Ellis was one of his executors), Ellis edited or revised the proofs of many of the Kelmscott Press productions, including Ruskin's *The Nature of Gothic: A Chapter of The Stones of Venice* and Geoffrey Chaucer's *Canterbury Tales*.[90] Chaucer was Morris' and Burne-Jones' favourite poet; the Kelmscott Chaucer, with eighty-seven woodcut illustrations by Burne-Jones, embodies Morris's ambition to reform book production based on medieval models.[91] Ellis was planning to publish an edition of Morris's *The Earthly Paradise*, with illustrations by Burne-Jones, though this was never to be realized.[92]

Frederick Ellis was a long-standing correspondent of John Ruskin (1819–1900). A volume of published letters written by Ruskin to Ellis, though incomplete, is ample testament to their growing rapport. Published

86   For example, copy letters from Ellis and Elvey to John Cole Nicholl; letters dated 21 April and 6 May 1887; Coll 425 Bx22 f7; dated 2 and 5 October 1888; Coll 425 Bx23 f8; dated 19 and 22 October, 4 December 1889; Coll 425 Bx50 f9; dated 27 January 1890; Coll 425 Bx50 f90. On Nicholl's collecting, see Murray McLaggan, 'The Library at Merthyr Mawr; A Bibliomaniac Great-Grandfather' <https://www.peoplescollection.wales/sites/default/files/chs0656oSirJohnNichollandsonofMerthyrMawr_0.pdf> accessed 24 August 2022.
87   Smith and Benger, *The Oldest London Bookshop*, 66.
88   'Their interests were identical'; Smith and Benger, *The Oldest London Bookshop*, 67.
89   Harris, 'William Morris and the Middle Ages'.
90   Bryan C. Keene, 'Cutting up the Middle Ages, Collecting the Renaissance: Illuminated Manuscripts in the Nineteenth Century', in Buron (ed.), *Truth and Beauty*, 217–20, here 218; Smith and Benger, *The Oldest London Bookshop*, 68.
91   Ellis, Osborne and Barringer (eds), *Victorian Radicals*, Cat. 81, 210–1.
92   Letter 12 of 1874 from Ruskin to Ellis; Ruskin, Stray Letters, 18.

in 1892, the collection gives a taste of their relationship, which had begun at the latest in 1870, when Ruskin wrote from Corpus Christi, Oxford asking Ellis to locate and purchase for him a recent translation of Giorgio Vasari's seminal *Lives of the Artists*.[93] Signing off on that occasion as 'Ever truly yours, J. Ruskin', his correspondence – although at first restricted to manuscript and book acquisitions – becomes increasingly warm and personal in tone, including an invitation to his home of Brantwood on the shore of Coniston Water in the New Year of 1874, and closing one of the last published letters in 1884 with 'Your loving J.R.'.. From Brantwood in 1883, following an illness, Ruskin muses:

> I am very happy [...] in the loan of your lovely Missal, – very happy in being able to covet missals, and take pride in my own work, once more. And very happy shall I be when I can shake hands again in that delightful library and chat-room of yours.[94]

This library was the inner room behind the shop at 29 New Bond Street (Figure 6.4). It is shown to have been furnished with glass-fronted and revolving bookcases, a large, cluttered desk, and Turkish rugs. The three-legged seat is positioned so as to present its occupant with a view of the shop through a raised flap in the intervening door. The shop and 'chat-room' at 29 New Bond Street appears to have been a nexus, a point of connection, at which all of these threads – people, bibliophilia, intellectual pursuits, artistic tastes and sensibilities – came together, through the person (in those days) of Frederick Ellis.

During the 1870s, Frederick had invited into the company his nephew, Gilbert Ifold Ellis (1858–1902). With publishing experience at Chatto and Windus, Gilbert soon established a strong reputation in his own right. Following his uncle's retirement in 1885, Gilbert set up in partnership two years later with Robert Elvey, Francis Nichols' nephew. They furthered Frederick's acquisition activities in a period that saw the dispersal of many fine libraries at auction, including that of William Morris.[95]

---

93 Letter 1, dated 17 February 1870; Ruskin, *Stray Letters*, 3.
94 Letter 38, dated 7 July 1883; Ruskin, *Stray Letters*, 63–4.
95 Smith and Benger, *The Oldest London Bookshop*, 70–1.

Figure 6.4: The 'library and chat-room' behind Ellis's Bookshop at 29, New Bond Street. Mezzotint of a photograph, courtesy of Gilbert I. Ellis and Robert V. Elvey, *Ellis and Elvey's General Catalogue of Old and Rare Books and Manuscripts* (London: Ellis and Elvey, 1894), title page.

In steadily drawing towards a conclusion, we return to address the research aim of Part Two of this chapter, namely, to understand the choice of the *Mirabilia urbis Romae* for translation by Francis Nichols and publication by Ellis and Elvey in 1889, bearing in mind the fallibility of its text as demonstrated by intervening academic scholarship. The publishers shared the Nichols' and the learned societies' commitment to making available in printed form rare or ancient texts and to disseminating information about where original manuscripts may be found.[96] Ellis and Elvey found themselves, moreover, at the heart of a community of writers, poets, artists, designers and craftsmen who saw the medieval past as a route to spiritual renewal and who revelled in medieval storytelling. *The Marvels of Rome* was not published for its scholarly accuracy about classical monuments, but it does contain colourful snatches of medieval myths – not about chivalric Knights of the Round Table or damsels in distress – but about the Christian holy figures themselves. An English translation of the *Mirabilia* therefore fed the yearning for 'high church' values spawned by the Oxford Movement, as well as the Pre-Raphaelite and Arts and Crafts predilection for the culture of the Middle Ages.

A handwritten copy letter dated 12 March 1889 from Robert Elvey to Messrs Nichols and Sons, who were printing the book, reveals that the market responded promptly to word that *The Marvels of Rome* was available: 'Have you any of the Mirabilia in? I have a few orders yesterday and today as it was announced as ready in last *Athenaeum*.'[97] The following day the request was repeated with more urgency: 'Can you get a few copies of the Mirabilia delivered to us this afternoon so that we can send these leaving by post today?'[98] The translator and editor of *The Marvels of Rome*, Francis Nichols, writing to his nephew, Robert Elvey, in November

96 To cite just one example of many, *The Letter in Spanish of Christopher Columbus, Written on his Return from his First Voyage, and Addressed to Luis de Sant Angel 15 Feb– 14 March 1493, Announcing the Discovery of the New World, Reproduced in Facsimile from a Unique Copy in the Possession of the Publishers* (London: Ellis and Elvey, 1889).
97 UCLA Coll 425 Bx23 f9.
98 UCLA Coll 425 Bx23 f9, written transversely across the page.

of the same year from his holiday address on the Isle of Wight, debates the merits or otherwise of their advertising strategy and hints at a *soupçon* of disappointment about foreign sales: 'I had hoped my book might have sold a little in America. Has it been sent to or noticed by any American Review?'[99] His trepidation was unwarranted. New York's *The Nation* recorded that the book was, 'Admirably printed, and bears every mark of competent scholarship',[100] while *The American Journal of Archaeology and of the History of Fine Arts* had been more expansive:

> Mr. Nichols has been for some years known as a very careful and accurate student of the topography of ancient Rome. The present charming volume is a popular contribution, and is addressed to a large public interested in the history of the eternal city and the vicissitudes of its monuments during the Middle Ages. [...] Many of the ancient buildings it describes have since been destroyed, hence even its scanty words are precious though slight indications. [...] Translations are given also of supplementary documents under five heads. [...] Mr. Nichols adds, to every part of these various texts, careful notes elucidating satisfactorily all the difficult points in topography, legend, or description, and showing a complete mastery of his subject.[101]

Reviews in the British press were equally effusive:

> Many people will be grateful to Mr. Nichols for his very careful English rendering of this curious and interesting little work. The translator's copious notes give the book a strong additional interest, and even an original value of its own.[102]

The 'supplementary documents' mentioned in the American review – additional material, over and above the original text of the *Mirabilia* – comprise five further elements: a description of Roman churches compiled in 1375;

99 UCLA Coll 425 Bx6 f6.
100 Gilbert I. Ellis and Robert V. Elvey, *Ellis and Elvey's General Catalogue of Old and Rare Books and Manuscripts* (London: Ellis & Elvey, 1894), unpaginated, inside back cover.
101 A. L. F., Jr, 'Reviewed Work: The Marvels of Roma [Mirabilia urbis Romae], or a Picture of the Golden City by F. M. Nichols', *The American Journal of Archaeology and of the History of Fine Arts* 5/3 (1889), 352–3.
102 *The Scots Observer*, as quoted in Ellis and Elvey, *Ellis and Elvey's General Catalogue*, unpaginated, inside back cover.

a description of Rome taken from the itinerary of a traveller, Benjamin of Tuleda, *c.* 1170; extracts from the *Ordo Romanus*, a treatise on the religious ceremonial of the papal court written by Canon Benedict (author of the *Mirabilia*); two papal bulls and the list of relics of the Lateran basilica, and finally a bird's eye view of Rome drawn in *c.* 1475 (including an enlarged detail of the Vatican) and Nichols' detailed description of the same.[103] Collectively these appendices are entitled by Nichols 'Mirabiliana'. As the reviews indicate, these passages are liberally annotated and painstakingly cross-referenced, underscoring what the American reviewer terms Nichols' 'complete mastery of his subject'.

Conclusion

In his description of the latter fifteenth-century bird's eye view of Rome, Nichols notes:

> The period to which our plan belongs is therefore precisely that which witnessed the commencement of the more critical studies of classical literature and epigraphy by which the authority of the *Mirabilia* was overthrown.[104]

This makes it abundantly clear that Nichols was under no illusion about the reliability, or otherwise, of his source. Let us then summarize the various motivations amongst the parties involved in its translation, publication and dissemination. Francis Morgan Nichols was proficient in the romance languages and a scholar of the classics and the art and antiquity of ancient Rome. As an expert on Roman archaeology and topography, he would have found it necessary for his own academic specialism to have translated from Latin into English the *Mirabilia urbis Romae* and the related texts contained in his 'Mirabiliana'. As Julian Pooley has shown in his many publications about the Nichols family between 1777 and 1873, through

---

103   Nichols (ed. and trans.), *Marvels of Rome*, 121–98.
104   Nichols (ed. and trans.), *Marvels of Rome*, 188.

generations they had demonstrated a strong commitment to making available in print unique, historic documents and rare texts. With this in his blood, it was but a short step for Francis to edit and annotate his own translations, have them published by a trusted publishing house and printed by his own family's press. Given his nephew Robert Elvey's new partnership with Gilbert Ellis, it was natural and logical for Francis to put publishing business in the direction of Ellis and Elvey.

On the other side of the coin, Elvey, wishing to make a favourable impression in his still-youthful partnership with Ellis, would have been keen to promote his uncle's scholarship and reputation as an expert in his field, at the same time as bringing trade into the firm. Working outwards from Robert Elvey, the firm – of Ellis (under Frederick Startridge Ellis) and Ellis and Elvey (under Gilbert Ifold Ellis) – was inextricably intertwined with some of the main protagonists of the Pre-Raphaelite, Arts and Crafts and Aesthetic movements, amongst whom medievalism flourished. The Ellises' close collaboration with John Ruskin and William Morris in particular brought them face to face with the Gothic Revival style and its inherent links to quasi-Catholic spirituality. While the research underpinning this chapter has revolved largely around the publication of a single volume within a restricted range of dates, it has spawned a number of potential, unplumbed depths, barely scratching the surface of the Ellis archive at UCLA Library Special Collections, Los Angeles.[105] Similarly, the Nichols Archive Database is fertile ground for further research.[106]

105   Ellis Booksellers Records, 1860–1928 (Collection 425), UCLA Catalogue Record ID 1217959. The collection comprises 11.5 linear feet of 23 boxes and 38 oversize boxes, mostly concerned with the sale of rare books; this was a gift to the library of George Smith made in 1951. The collection is arranged in the following series: Correspondence (Boxes 1–11), Letter books (Boxes 12–50), Trade day books (Boxes 51–3), Day books (Boxes 53–61). Correspondents include Dante Gabriel Rossetti, William Morris, Sir Lawrence Alma-Tadema, Sir Edward Coley Burne-Jones, John Maynard Keynes, George Bernard Shaw and Joseph Crawhall (one of the 'Glasgow Boys').

106   The Nichols Archive Database is accessible by appointment with Julian Pooley, Public Services and Engagement Manager, at Surrey History Centre; <https://www.surreycc.gov.uk/culture-and-leisure/history-centre/visit> accessed 10 August 2022.

As for the *Mirabilia urbis Romae*, written with the idiosyncratic quirkiness of a pre-humanistic text, its English translation as *The Marvels of Rome*, liberally corrected and explained with erudite notes and appendices, had its own niche markets amongst those interested in classical and medieval Rome, the early martyr saints and the Middle Ages in their own right.

ANTHONY QUINN

7 A Whirlwind Tour of Tourism in Magazines: 1851 to 2020

In 1841, Thomas Cook hired a train and took 485 people on a twelve-mile train trip in the East Midlands of England from Leicester to Loughborough – the world's first package holiday and the advent of modern mass-market tourism.[1] He then set up a bookselling and printing business, and arranged trips for temperance societies and Sunday schools. Although these excursions laid the foundations of his future business, Cook made little money from them at first.[2] Forty years later, the Manchester entrepreneur George Newnes launched *Tit-Bits* and then *The Strand*, establishing a model for mass-market periodicals. Ever since, the tourism and magazine industries have grown hand-in-hand: advertising revenue from the former encouraged magazines to promote travel to their readers. To complete the virtuous circle, specialized tourism spawned ever-more-niche magazines. However, as Steward has argued, 'not enough credit has been given to the role of the press as a promoter of tourism.'[3]

In the early nineteenth century, the grand tour was a spur for literary endeavours among aristocrats. From the 1830s, the expanding press

1   However, John K. Walton discusses special trains running in 1836: 'Thomas Cook: image and reality' in Richard W. Butler and Roslyn Russell, eds, *Giants of Tourism* (London: CABI, 2010), 84. Also see F. Robert Hunter, 'Tourism and empire: the Thomas Cook & Son enterprise on the Nile, 1868–1914', *Middle Eastern Studies*, 40/5 (2004), 28–54.
2   'Leicester – the birthplace of popular tourism', Leicester City Council <https://www.storyofleicester.info/city-stories/thomas-cooks-leicester/> accessed 6 October 2025.
3   Jill Steward, '"How and where to go": the role of travel journalism in Britain and the evolution of foreign tourism, 1840–1914' in John K. Walton (ed) *Histories of Tourism* (Bristol: Channel View, 2005), 40.

encouraged readers to travel. Furthermore, publishers held up a mirror in which different kinds of tourists were able to see the tastes and preferences of their social circle depicted. In the second half of the century, tourism for Britons became an 'expression of distinctive personal and social identities' as well as a global industry.[4] Wealth was spreading as Britain's population and economy expanded. Literacy levels rose. Thomas Cook discovered a ready market of potential travellers and readers. He launched his own magazine, *Cook's Excursionist*, in 1851, in which he posed the question: 'How are working men [and families] to get to the [Great] Exhibition?' He provided the answer, taking 150,000 people to the spectacle in London's Hyde Park. Four years later, Cook ran a 'grand circular tour' of Belgium, Germany and France. His son, John Mason Cook, after working for the Midland Railway and running a printing firm, built on his father's work to create an international travel system.[5] By the end of the century, the company had taken 12,000 people to Palestine alone, including the German emperor Wilhelm II, an event marked with a photograph on the cover of *The Quiver* in November 1898.[6]

Yet Cook's tours were ridiculed by the artists and writers of the humorous weekly *Punch*. Among these was Richard Doyle, who drew the cover design that *Punch* used for a century. From 1849, his *Manners and Customs of Ye Englyshe* cartoons ridiculed the excursionists. Later, *Punch* co-founder Henry Mayhew and the caricaturist George Cruikshank combined to deride trips to the Great Exhibition with their Sandboys family characters. Doyle continued the theme and in 1854 he took a trio he had developed in *Cornhill Magazine* and *Punch* on a trip round Europe for a book, *The Foreign Tour of Messrs Brown, Jones and Robinson*. Walton refers to such 'snobbish reactions' in the London press.[7] Steward points to magazine attacks against Cook's 'Cockney hordes' being taken up by

4 Steward, 'How and where to go', 39.
5 Walton, 'Thomas Cook: image and reality', 89.
6 Hasan Ali Polata and Aytuğ Arslan, 'The rise of popular tourism in the Holy Land', *Tourism Management*, 75/12 (2019), 231–44.
7 Walton, 'Thomas Cook: image and reality', 89.

upmarket newspapers.[8] In reply, Cook denounced the 'hirelings and witlings of a very small section of the London press' in the *Excursionist*. In an 1864 article in *All the Year Round*, which was owned by Charles Dickens, Cook defended his parties as including 'ushers and governesses, practical people from the provinces, and representatives of the better style of the London mercantile community'. Press coverage improved.[9] In 1879, the journalist George Augustus Sala wrote: 'Mr Cook can afford to smile at his detractors.'[10]

The success of the *Excursionist* led Cook to launch editions in New York and Bombay. In 1867, these had a circulation of 58,000; by 1892, it was 120,000 from six international editions.[11] A 1904 *Pearson's Magazine* article portrayed Thomas Cook & Son as 'The patron saints of modern travel'.[12] In 1928, the founder's grandsons sold the business, yet the *Lady's Companion* still reported on births in the family in 1933.[13] Cook's was not without its rivals, however. A 1900 issue of *The Butterfly* carried an advert for tours organized by Henry S. Lunn.[14] Like Cook, Lunn mixed religion, travel and publishing. In the 1890s he led trips to religious conferences in Switzerland. He was also editor of *The Review of the Churches*, a periodical published from 1891 to 1896. All this spawned a business based on winter vacations and religious tours to Italy and the Holy Land. Combining travel and activity 'revolutionized the leisure travel industry'.[15] Sixty year later, Lunn Poly was one of the largest travel agents.

8   Steward, 'How and where to go', 45–6.
9   Walton, 'Thomas Cook: image and reality', 87–8.
10   Steward, 46–7.
11   Steward, 42–6.
12   Marcus Woodward, 'The patron saints of modern travel', *Pearson's* (May 1904), 517–24.
13   Walton, 90. *Lady's Companion*, 'New babies born to the family of Thomas Cook' (21 October 1933), 1.
14   *The Butterfly* (January 1900), inside front cover.
15   'Lunn Poly', Oxford Reference <https://www.oxfordreference.com/view/10.1093/oi/authority.20110803100119954> accessed 26 May 2022.

Royal Influence

From her childhood, Queen Victoria took holidays on the Isle of Wight off England's south coast and her wealthy subjects followed. Her home there inspired the name of a monthly magazine, *Osborne*, in 1896. This followed Ward, Lock & Co.'s *Windsor Magazine* the previous year, and *Balmoral Magazine* was founded in Aberdeen in 1903, to mark a trio of royal residences. Victoria also highlighted Scotland in 1867, when *Cornhill Magazine* published her Highlands journals. More aristocrats and the middle classes took to writing articles. The June 1888 issue of *Murray's* included 'A lady's winter holiday in Ireland' by Isabella Bird, the first woman fellow of the Royal Geographical Society. The Countess of Meath wrote 'A woman's thoughts on travel' for *The Quiver* of December 1890.

Alongside the articles were travel columns. An early example was in *The Queen*. This had been founded in 1861 by Samuel Beeton, 'one of the first editors to grasp that writing about tourism could help sell magazines'.[16] The column answered queries on topics such as the best routes to resorts.[17] By 1922, there was an Automobile World column, which also covered motor camping and caravanning; Travel Notes; and answers to correspondents.[18] Among the advertisers was Instone Air Line, with a daily service to Paris and Brussels. The first issue of *Ladies' Field* in 1898 carried a Travel Talk column, in which Sir Edward Sieveking, Victoria's physician, advocated travel for everyone, though he deprecated 'the lack of knowledge we have of our native land, as a result of the zest for going abroad'.[19] A decade later, that 'zest' appears to have galvanized the *Gentleman's Journal*. It urged readers: 'Become a member of our Home-Touring Coterie ... and thereby promote Home Travel and the fuller knowledge and enjoyment of your Own Country.'[20]

16 Steward, 42–5.
17 *The Queen*, 28 January 1893, 150 & 153.
18 *The Queen*, 22 September 1922, pages 308, 309, 311 and 313 respectively.
19 *Ladies' Field* (19 March 1898), 43.
20 *Gentleman's Journal* (21 August 1909), 645.

Tourism came to identify patterns of consumption and lifestyles, so publishers chose their featured destinations and type of holiday to appeal to a certain readership. *The Royal*, a rival to *The Strand*, showed women on a gondola in Venice; the women's weekly *Home Chat* ran a coupon for children's balloons at resorts while rival *Home Notes* offered prizes for the best shells; and *The Strand* concentrated on August holiday fiction from 1914.[21] Travel themes also appeared in advertising. Beecham's Pills produced a 'holiday alphabet' that included 'J is for journeys by sea, lake and land'; 'W for wheeling, which thousands love well'; and 'Z for the zoo, for young folks just the thing'.[22]

Transport Technologies

Each new transport technology spawned magazines, with a boom for cycling titles from 1875, motoring from 1895, railways from 1897 and flying from 1909. The Bicycle Touring Club was founded in August 1878 and issued a monthly circular from October.[23] Publishers had already seen the trend, with *Ixion*, *The Bicyclist*, *Bicycling News* and *The Cyclist* among the pioneers.[24] In the 1880s, a dozen titles were available. *Cycling* from 1891 was the foundation of Temple Press, a transport publishing group, which competed with *Bicycling News* and *The Cyclist* from Iliffe publishing based in Coventry, a centre for manufacturing.[25] Alfred Harmsworth – later Lord Northcliffe – was editor of *Bicycling News* before founding *Answers* and

---

21 *The Royal* (June 1900), front cover; *Home Chat* (31 July 1909); *Home Notes* (1 August 1914), 215.

22 *Forget-Me-Not* (28 September 1895), back cover.

23 Sheila Hanlon, 'The history of Cycle magazine' (2017) <https://www.cyclinguk. org/cycle/history-cycling-uk-magazines-beginnings-victorian-golden-age-cycling-press-1878-1900> accessed 6 October 2025.

24 Reginald Pound and Geoffrey Harmsworth, *Northcliffe* (London: Cassell, 1959), 57.

25 See Arthur C. Armstrong, *Bouverie Street to Bowling Green Lane* (London: Hodder & Stoughton 1946) for a history of Temple Press.

the *Daily Mail*.[26] In 1896, launches included *The Hub*, 'for wheel-men and women', from George Newnes. Two other titles were *Cycling World Illustrated* and *Lady Cyclist*.[27]

The expanding rail network created an audience of millions of commuters and travellers for publishers – and a national distribution system. Furthermore, stations provided stalls for stationers in Britain and overseas. In 1900, *The King*, a news weekly, was sold at railway and other bookstalls; outlets in Paris, New York, Australia and South Africa; and the main stations of India.[28] According to the back page advert in the *Cornhill Magazine* of October 1876, a million pounds had been paid out for deaths and injuries by the Railway Passengers' Assurance Company. Readers' fear of accidents inspired *Tit-Bits*, one of three popular weeklies: 'A current copy of *Tit-Bits* is a free railway accident insurance policy for £100,' covers declared. By November 1888, £1,500 in claims had been paid. *Answers* offered £500 policies[29] and described itself as 'for home and train' into the 1930s.[30] The third of the big weekly rivals, *Pearson's Weekly*, raised the offer to £1,000 in 1904 (15 December), a figure it kept into the 1930s.[31] The Great Western Railway had its own *GWR Magazine* in 1888. *Railway Magazine* launched in July 1897. *Locomotive* was another competitor. *Railway and Travel* came out in 1910 and *Railways of the World* in 1924. *LMS Railway Magazine* was published by the London, Midland and Scottish Railway from 1923.

From the mid-1890s, magazines switched from standard front covers to illustrations that changed with each issue or sought revenue from semi-display advertising. They also began to use colour.[32] Some showed themselves being read on trains. The *Penny Magazine* has a cover of itself

26 Pound & Harmsworth, *Northcliffe*, 55.
27 'Romantic freedom and the bicycle', University of Warwick <https://warwick. ac.uk/services/library/mrc/explorefurther/images/newwoman/romance> accessed 6 October 2025.
28 *The King* (1 September 1900), 283.
29 *Answers* cover, 14 June 1890.
30 *Answers* issues of 16 May 1891, 6 November 1909, 3 January 1925, and 24 September 1932.
31 *Pearson's Weekly* cover, 4 April 1931.
32 Anthony Quinn, *A History of British Magazine Design* (London: V&A, 2016), 46–103.

and other Cassell titles on a station stall in 1904.[33] Another example is a self-referential cover of *Pall Mall Magazine*: a woman is waving from a train with a copy of the magazine on which she is depicted (June 1910). The back cover of a 1936 issue of *The Sphere* (1 February) was a house advert from the publishers, Illustrated Newspapers. Above the selling line 'Good travelling companions', a couple in a first-class compartment have copies of *Illustrated London News*, *The Sketch*, *Bystander*, *The Tatler*, *Sporting and Dramatic News* and *The Sphere* itself.

Pearson's of August 1905 was a special railway number. To mark the centenary of rail in 1925, *Punch* produced a supplement with a train-based parody of Doyle's long-standing cover. In its early years, *Punch* 'was far more concerned with emphasizing the dangers and discomforts of the new mode of travel and the wild speculation that led to it'. In contrast, the supplement showed queues at that year's British Empire Exhibition to see the *Flying Scotsman* locomotive, 'every boy's dream'.[34]

As the car industry developed, *Autocar* (Iliffe) launched in 1895 'in the interests of the mechanically-propelled road carriage'. It was followed between 1902 and 1909 by *Car Illustrated*; *Motorcycling and Motoring*, which became *The Motor* (Temple Press); *Car Magazine*; *Motor Cycle*; and *Motorist and Sportsman*, 'plain chatty English ... will be its motto'. Both *Car Illustrated* and *Car Magazine* were edited by the future Lord Montagu. His son founded Beaulieu National Motor Museum. Later titles specialized, but all encouraged motoring holidays. Launches included *Cycle Car*, *Light Car*, *Brooklands Gazette* ('racing and sporting events'), *Austin Magazine*, *Motor Sport*, *Speed* (British Racing Drivers' Club), *Morris Owner*, *Modern Travel* (Sunbeam and Talbot owners), *Bugantics* (Bugatti), *Practical Motorist*, *Modern Travel* (Sunbeam and Talbot), the *Scooter* and *Top Gear* (1955).

Motoring became a fashionable activity. A 1906 frontispiece for *The Bystander* was a photograph of Victoria Godwin, 'a brilliant feminist motorist' racing a 40hp Ariel motorbike. *Home Chat* of 2 July

---

33   Anthony Quinn, 'The self-referential magazine cover' <https://magforum.wordpress.com/2015/04/01/the-self-referential-magazine-cover> accessed 6 October 2025.
34   'The railway centenary', *Punch* supplement (1 July 1925), 1.

1920 suggested motorcycle sidecar holidays. In 1926, Newnes started its *Motorists' Touring Guide* with a car as a crossword prize. The January 1931 cover of *Morris Owner* showed a woman driving in the Alps.

Another form of transport was still in its infancy – the aeroplane. In 1903, the Wright brothers made their four-mile flight and three years later Lord Northcliffe offered £10,000 in the *Daily Mail* for the first flight from London to Manchester. In 1907, the *Mail* sponsored the first model aircraft prize show with 129 entrants.[35] *Tit Whits* satirized the event:[36]

> The *Daily Fail* in a stormy high
> Did get some men to try and fly.
> The horns did blow, but the things won't go
> And what was the good no one does know.

The winning model was by Alliott Verdon Roe, who set up the Avro aircraft maker three years later.[37] The prize for Northcliffe's 160-mile Manchester challenge was not claimed until 1910. *Flight* was launched in 1909 by the Royal Aero Club as the first such weekly in the world. It was devoted to 'the interests, practice, and progress of aerial locomotion and transport'. *The Aero* from Iliffe came out the same year. These were followed by *The Aeroplane* and *Flying*.

Many of the aircraft that flew during World War I were built by Airco, a company founded in 1912 by George Holt Thomas. His father owned *The Graphic* and *Daily Graphic*, and George had made his own fortune by founding *The Bystander* and *Empire Illustrated*. In 1919, an Airco subsidiary, Aircraft Transport and Travel, began the first passenger service from London to Paris.[38] Five years later, Imperial Airways was formed. These innovations encouraged the forming of Airways Publications in 1924, which launched *Airways*.[39] Competition landed in 1927 with another monthly, *Air*, from the Air League of the British Empire, which promoted

35 'Machines by which man may be able to fly', *The Sphere* (13 April 1907), 32.
36 'The flying machine', *Tit Whits* (May 1907), 4 & 5.
37 'British Civil Aviation in 1907', RAF Museum <https://www.rafmuseum.org.uk/research/research-enquiries/history-of-aviation-timeline/british-civil-aviation/1907-2/> accessed 6 October 2025.
38 *Anchor Magazine* (1931), 102.
39 *Airways* (November 1926), front cover.

'air-mindedness'. Such developments fascinated general magazines as well, particularly the society weeklies and monthlies. *The Bystander* carried a Flying Notes column, which reported in 1929 on 'air taxis' from Hanworth Park in North London at 'a shilling a mile' (22 May). The 1930s saw the arrival of *Air and Airways*, and *Airways and Airports*. *Popular Flying*, aimed at the general public, was launched with W. E. Johns as its editor (it published his popular Biggles air adventure stories). The cover of June 1937 promoted 'aerodrome holidays' over Britain. However, flying was the preserve of the elite. The lot of most people was captured by 'Your Country Needs You' poster artist Alfred Leete for a *London Opinion* cover.[40] He portrays the holidaymaker as a suitcase-carrying fish tempted by landladies with 'board residence' and 'apartments' as their bait.[41]

Leisure Magazines and Holiday Specials

Alfred Harmsworth sought to broaden the cycling sector in 1897 with *The Rambler*, a penny weekly 'devoted to the out-door life'. *Cycle Camping* was the organ of the Association of Cycle Campers (1906) and *Camping* was for the Amateur Camping Club (1907). George Newnes launched *The Wide World* in April 1898, full of 'true narrative: adventure, travel, customs and sport'.[42] By 1901, *The Times* newspaper reported that the 'demand for holidays [was led by] cheap and easy locomotion and greater prosperity'.[43] The year after, *Picture Postcard Magazine* began catering for collectors. In 1911, *The Manxman*, 'the three-legged magazine', promoted the Isle of Man as 'the holiday pivot of the empire'. Before 1914, the working week was fifty-five hours, including four hours on Saturday mornings. The half-day Saturday had inspired the name of *Ally Sloper's Half Holiday*, a comic-strip

40 Anthony Quinn, '*London Opinion* – the most influential cover', Magforum <http://www.magforum.com/mens/london-opinion.htm> accessed 6 October 2025.

41 Alfred Leete, 'Bait!', *London Opinion* (20 June 1925), 367.

42 Advert for *Wide World* in *King of Illustrated News* (1 September 1900), 286.

43 'The demand for holidays', *Times* (16 September 1901), 7.

magazine for adults from 1884. By 1919 the working week was forty-eight hours.[44]

After the Great War, the Cunard Steam Ship Company launched *The Cunarder*.[45] It carried general articles, such as visiting land-locked Switzerland, as well as material about its liners. The *Blue Peter*, a magazine of sea travel, promoted the P&O line in its colour supplements from 1921. However, it was a time of recession and the 1920 holiday number of *Home Chat* depicted a dozen cheaper activities such as camping, a horse-drawn wagon, motorcycle and sidecar, reading on a beach, taking photos, walking and picnics.

Risqué magazines such as *Photo Bits* had added drawings and photographs of women in bathing suits to their dancing girls and music hall stars since the turn of the century.[46] A 1911 issue showed a woman riding a sea serpent above a poem, 'The Flight to the South':

> Away from the fog of London town.
> Away in a beautiful bathing gown.
> Away to the place where the sun shines bright,
> And spooning is free in the clear moonlight![47]

Winter sports numbers were becoming established. These often showed women skiing, including *The Royal* (January 1926), *Eve* (10 November 1926) and *The Bystander* (7 November 1928). With the Great Depression, walking became popular, and *Hiker & Camper* made its debut in February 1931. It also discussed motor-caravans.

People on holiday have time for reading, so publishers produced special numbers. An early example is the 1877 seaside number of *Family Herald* (no. 1787). It cost two pennies, twice the normal weekly issue. A special cover masthead showed: children on a beach; a couple under a parasol above a bay; fishermen; bathing huts; and a seaside town complete with lighthouse. The cover of the 1898 'grand double seaside number' of

---

44 D. McKitterick, *A History of Cambridge University Press*, vol. 3, (Cambridge: Cambridge University Press, 2004), 361.
45 *Cunarder* (September 1926), 12–4.
46 See *Photo Bits*, 9 June 1906, 8–9; 25 June 1910, cover; and 6 June 1909, 15–6.
47 *Photo Bits* (7 January 1911), 11.

Woman's Life (July 2) showed a child by a bathing hut with her bucket and spade.

The trend for specials continued, particularly at Easter, Whitsuntide, in August, and at Christmas; the *Illustrated London News* had pioneered colour supplements at Christmas since 1855.[48] Examples of special issues included *Answers* (5 May 1891); the cheap weekly *Snap-Shots* (6 April 1901); the upmarket *Pall Mall* (June 1910); fiction monthly *Cassell's* (August 1928); and *The London* (August 1929). *Punch* published a sixpenny summer number in 1916 (5 July). The colour summer cover became a tradition, even during World War II rationing. Specials for monthly magazines usually doubled the cover price, whereas weeklies produced premium extra issues. *London Opinion*'s 1924 Christmas special cost six pennies – three times the usual price – and the cover was in three colours; in 1927, it was four colours. In 1929, there were specials for Easter (*6d.*, April 6) and Whitsun (*3d.*, May 25) and summer and Christmas extras (both *6d.*). *Pictorial Weekly* of 16 April 1927 was an Easter number, 'full of holiday reading'. Among the articles was 'Good rooms to let' in which a seaside landlady 'talks about the £sd [pounds, shillings and pence] of a boarding house'.

In 1934, Sir Evelyn Wrench argued in *Everyman*, a news weekly, for ten days' holiday for every worker. Wrench had made his name producing picture postcards, founded the English-Speaking Union, and owned *The Spectator*.[49] In the same year, *Pictorial Weekly* discussed the 'Continental holiday craze', while the travel writer S. P. B. Mais identified the loveliest English villages. The back cover was an advert for the Automobile Association motoring club.[50] A 1934 *John Bull* headline forecast that 'Nothing can stop air travel', not even air crashes.[51] The 1936 summer number of *John O'London's Weekly* added a gravure section entitled 'Traveller's joys'.[52] An advert for the London School of Journalism promoted its £50 [Lord] Northcliffe prize for holiday journalism. The Christmas special

48 Quinn, *British Magazine Design*, 46.
49 'Every worker, says Sir Evelyn Wrench, should have an annual holiday on full pay,' *Everyman* (23 February 1934), 9.
50 *Pictorial Weekly* (4 July 1934).
51 J. Stubbs-Walker, *John Bull* (13 October 1934).
52 *John O'London's Weekly* (20 June 1936), i–viii.

used a Kodak snapshot as its frontispiece. Kodak had linked the camera with holidays since at least 1901 and magazines encouraged readers to send in their snaps for publishing.[53] One campaign with a £1,000 prize inspired fiction in a 1912 *Home Chat*, 'Cupid and the Kodak: a holiday story', by Berta Ruck, who would go on to write scores of romance novels.[54] The company also produced its own magazine from the 1920s.

Mainstream magazines were starting to experiment with printing colour photographic covers. *Vogue* chose a woman in a red bathing suit with a beachball against a blue sky for its first example and the subject for *Home Notes* was a couple on a beach.[55] It was a costly and lengthy process and did not become common until the 1950s. The editors of *Woman's Own* were more concerned about the perils of summer and romance, with the cover lines 'Sunbathe – but don't spoil your looks' and 'Should they holiday together?'[56] A 1938 cover of *Woman* had a partial solution – sunglasses.[57] Some products were promoted in new ways. The 1929 summer number of glossy weekly *Piccadilly* carried three examples. Pond's cream promised protection from sun and wind. An enterprising chemist offered toiletries to 'banish ... sunscorch, tan, freckles, enlarged pores, crows-feet and sun-faded hair'. Wright's soap gave protection from infection and was refreshing after a day at the races. Of the issue's ninety-six pages, twenty-nine were advertising and ten were travel-related. But *Piccadilly* went further: it ran a travel bureau, which was promoted on three pages. Twenty editorial pages covered travel.[58] An early press advertisement for Guinness was a quarter-page in *John Bull*, one of the best-selling weeklies in 1929. It stated: 'Guinness for holidays. Guinness is good for you.'[59] The brewer had only started running national

<hr>

53 *The Harmsworth* (August 1901), Kodak advert on contents page; Home Chat (15 July 1922), 46.
54 *Home Chat* (31 August 1912), 449.
55 *Vogue* (20 July 1932). *Home Notes* (15 August 1936).
56 *Woman's Own* (22 July 1935).
57 *Woman* (9 July 1938).
58 *Piccadilly* (15 June 1929).
59 *John Bull* (17 August 1919), 17.

advertising campaigns in February that year.[60] Elsewhere, Cow & Gate's milk powder was 'the perfect holiday food' for a baby on a beach.[61]

At the end of the 1890s, publishers began to explore the possibility of a dedicated consumer travel magazine. An early attempt was *The Tourist* from Sheffield's *Weekly Telegraph*, in 1897. Its January 1899 cover proclaimed: 'We English are fond of rude [i.e., simple] travel.' Articles were devoted to lady tourists; the Isle of Man in winter; news from Brazil; plans for the USA to admit bicycles belonging to members of the Cyclists' Touring Club for nothing; and a sketch of the latest biking costume.[62]

The following year, *The Traveller* – 'For whom the world is a playground' – was published by George Newnes with the co-operation of Thomas Cook & Son (Figure 7.1). However, it only lasted a couple of years. For the issue of 11 January 1902, advertising included Dunlop cycle tyres, the Hydro hotel in Church Stretton, Ilford photographic plates, and filters to avoid typhoid fever. Shropshire was 'The Highlands of England'. A factor in *The Traveller*'s demise may have been the sacking of the editor over a libel against Brunel's Great Steamship Company.[63]

In May 1919, *Holiday Resorts* made its debut for 'tourists, nature lovers, health seekers and pleasant reading'. *Outward Bound* appeared in 1920, with *Thirty-Nine Steps* author and keen birdwatcher John Buchan a contributor. *Open Air*, 'for lovers of nature and outdoor life', was a sister title to *Country Life*. London County Council Tramways published *The London Holidaymaker* as a free annual. Its 1933 cover encouraged Londoners to visit free sights in the capital. This was the time of the Great Depression. The magazine notes: 'If you must economize, no place in the world will give so much for nothing, so much "wheel" mileage for such few pence.'

60 First advert was on 6 February 1929 in the *Daily Express*.
61 *Practical Housekeeping*, 2 July 1938, 4.
62 *The Tourist* (January 1899).
63 *Times* (22 April 1902), 3.

Figure 7.1: *The Traveller* (11 January 1902) was published by George Newnes with the co-operation of Thomas Cook (author's collection).

Mass Air Travel

In 1949, Horizon Holidays took its first package charter to Corsica in a former military transport. Vladimir Raitz, a Russian émigré and once a Reuters correspondent, founded Horizon with the idea of offering an inclusive flight and hotel. He began by advertising in the *New Statesman* and *Nursing Mirror*.[64] Soon, a million Britons a year were flying abroad.[65] In 1950, the *Sunday Times* started publishing holiday guides. A year later it purchased *Go* travel magazine. However, the title was in the hands of a different publisher, Through Europe, by 1959. The cover destinations were Rome, lesser-known parts of France, and Italy's two Rivieras. Inside, more unusual fare was offered in advertising for Freddie Laker's Channel Air Bridge using car-carrying aircraft from Southend.[66] Among the advertising was a page for the Volkswagen Beetle as 'a superb mountain climber – an ideal holiday car'.

The new Queen Elizabeth's tours provided inspiration for holiday-makers. This was looked back on by 'Seventy years of royal globetrotting' in the *Sunday Times* for her platinum jubilee in 2022. The article identified 'lavish digs' that had hosted the monarch, from Kenya's Treetop Lodge in 1952 to a Berlin hotel in 2015.[67] Soon after her 1953 coronation, the Queen set out on a 44,000-mile Commonwealth tour. *Punch* ran special versions of its Doyle cover with Mr Punch and his dog Toby waving off the royals' Comet jet (25 November); six months later they welcomed back the new royal yacht, *Britannia* (12 May). Weeklies such as *Woman's Own*

---

64 Simon Ashley, 'Promotion keeps trade at the top of the league', *Travel Trade Gazette 40*th *Anniversary* (September 1993), 47, 49–50.
65 David Churchill, 'Wraps off for the holiday package', *Financial Times* (9 September 1989).
66 Page advert. 'Fly with your car from Southend', *Go* (April 1959), 3.
67 *Sunday Times*, travel section (29 May 2022), 14–5.

and *Woman*, which had a combined sale of 5.5 million in 1959, followed the royals closely.[68]

Travel Trade Gazette, 'the world's first weekly travel newspaper', appeared in 1953.[69] At the time, only 3 per cent of British holidaymakers went abroad.[70] A *TTG* editorial noted: 'Competition to capture new classes of travellers – those in the lower income groups and those who have never been abroad – is likely to produce a number of original experiments in low-price holidays.' For most Britons, the resorts of choice were Blackpool, Brighton and Bognor, alongside Billy Butlin's camps. The 1958 summer special of *Blighty*, a popular weekly, covered a Miss Holiday Belle contest in Ilfracombe. Also, the magazine sent 'Blighty Bill' to resorts to give premium bonds to people who carried a copy of *Blighty* and recognized him.[71]

As established airlines switched to jets from 1958, the cast-off aircraft enabled entrepreneurs to start charter services. Companies such as Clarksons, Global and Cosmos sprang up and Thomas Cook was never able to dominate jet holidays in the way it had rail.

Soon, for the middle classes, nothing less than a fortnight in Spain was enough.[72] The decade saw consolidation. Cook's took over rival agents Dean & Dawson, which dated from 1871 and whose press, Deanprint, still exists.[73] The 1956 issue of Thomas Cook's annual *Holidaymaking*, founded three years earlier, carried both companies' branding (Figure 7.2). It has

68 Anthony Quinn, 'Women's magazines: sales figures 1938–59', Magforum <http://www.magforum.com/glossies/womens_magazine_sales.htm>. Accessed 6 October 2025.

69 Anon, 'TTG and travel trade have come a long way together', *Travel Trade Gazette 40th Anniversary* (September 1993), 7.

70 *Daily Herald* survey quoted in 'Travelling through the years', *Travel Trade Gazette 40th Anniversary* (September 1993), 8.

71 *Blighty* (16 August 1958), 27.

72 Paul Smith (ed.), *The History of Tourism: Thomas Cook and the Origins of Leisure Travel* (New York: Psychology Press, 1998), 5.

73 'Deanprint Ltd – The first 125 years' <https://www.deanprint.co.uk/deanprint-ltd-first-125-years> accessed 6 October 2025.

Figure 7.2: Thomas Cook's *Holidaymaking* in 1956. It mimicked popular women's weeklies with text that was 'light, romantic and fluffy' (author's collection).

been described as a watershed because it rejected the dry style of bro-chures and earlier magazines. Instead, *Holidaymaking* took inspiration from women's weeklies.[74] The cover was a young woman on a beach with two men chatting her up. In retrospect, the text was seen as 'light, romantic and fluffy' by the *Travel Trade Gazette*, an approach that influenced travel journalism, particularly newspaper supplements in the 1960s when the copy was 'romantically written' to appeal to women because they 'made the decisions where to go'.[75] *Holidaymaking* promoted Spain's Costa Brava, a term dating back to 1908, and Costa del Sol, coined in the 1920s. In 1957, British European Airways started flying to Valencia and the term 'Costa Blanca' was invented for the coastline south of the city.[76] In a 1963 *Sunday Times Colour Magazine* feature, E. D. O'Brien, founder of a pub-lic relations agency, said he had invented the term Costa Blanca. One of his clients was Spain, and he pointed out that visitors had jumped from 13,000 a year when he won the account in 1949 to 800,000 in 1962.[77] The country entertained 1.5 million Britons in 1965. *Titbits* (the subtly renamed *Tit-Bits*) carried an eight-page holiday special in 1982, which talked about a price war among tour firms.[78] However, British visitors peaked in 1988 at 7.6 million, when sunseekers turned away from the Costas to Greece, Italy and the US state of Florida. A more unusual angle was hippy travel in *Oz*, the underground magazine. Alongside a 'Smack kills' spread, was 'Indian summer', an 'Oz package tour feature' about Goa.[79] In a later issue was 'Bali: hippy paradise jungle'.[80]

By 1993, 12.6 million Britons were taking an overseas holiday, a fifth of the population, with one in ten flying long-haul to Florida, the Caribbean or Thailand.[81]

<hr>

74 Simon Ashley, 'Promotion keeps trade at the top of the league', *Travel Trade Gazette 40th Anniversary*, (September 1993), 47, 49–50.

75 Lyndy Stout, 'Press gang takes a more mature approach to travel', *Travel Trade Gazette 40th Anniversary*, (September 1993), 52.

76 'Money goes further on the unspoilt Costa Brava', *Holidaymaking* (1956), 34. *Contours of Europe 1958*, Contours Ltd brochure, 12.

77 Richard West 'Paid to Persuade', *Sunday Times Magazine* (7 April 1963), 2–3.

78 Jeff Mills, 'Spain still reigns for value', *Titbits* holiday section (18 December 1982), cover.

79 *Oz* (September 1971), no 36, 14–5.

80 *Oz* (July/August 1972), no 43, 52–5.

81 Peter Lilley, 'Trade earns respect as it pays its way', *Travel Trade Gazette 40th Anniversary*, (September 1993), 60.

Travel and The Colour Supplement

A six-week print strike in 1959 established forty hours as the standard working week for most people.[82] Magazines and newspapers realized the advertising potential of the booming holiday industry and started travel supplements early in the year, a quiet time for advertising.[83] Elizabeth Nicholas, *Sunday Times* travel editor in 1960, has been described as creating the idea of holidays among the public in the same way as the cookery writer Elizabeth David introduced foreign food. However, according to a former travel editor of *The Times*, much journalism at the time was repetitive. 'We would go off on press trips and then you would get a crop of articles all about the same place ... There was very little criticism.'[84] So Nicholas and others founded the British Guild of Travel Writers, seeking a more professional attitude among journalists.

Ian Fleming captured the glamour of exotic locations with his Bond books from 1952. He had joined *The Times* in 1938 and then worked for Naval Intelligence. After the war, he ran the foreign desk at the *Sunday Times*. Fleming left the paper in 1959 but was involved with the launch of its *Colour Section* in February 1962.[85] 'The Living Daylights' was flagged on the supplement's cover as a 'new James Bond short story'. Later in the year *Dr No* was released, and 007 became a massive movie franchise.[86] The influence of Bond on travel can be seen in examples such as the men's monthly *King* referring to Istanbul, 'where one of the sights is the James Bond sewer' seen in *From Russia with Love*.[87]

82 McKitterick, *A History of Cambridge University Press*, 361.

83 John Ruler, 'Majestic characters and the birth of the guild' (23 September 2020) <https://bgtw.org/the-sixties> accessed 6 October 2025.

84 Lyndy Stout, 'Press gang takes a more mature approach to travel', *Travel Trade Gazette 40th Anniversary*, supplement, September 1993, 52.

85 The supplement was called *The Sunday Times Colour Section* for the first year. The issue dated 10 February 1963 became *The Sunday Times Colour Magazine*. See Quinn, *British Magazine Design*, 154.

86 Mark Edmonds, 'My secret life at the *Sunday Times*', *Sunday Times* (14 October 2012) <https://www.thetimes.co.uk/article/my-secret-life-at-the-sunday-times-s3lg-pxszfhz> accessed 6 October 2025.

87 'Getting away', *King* (March 1966), 95–8.

The supplement struggled in the first year, with issue sizes of twenty-four or thirty-two pages.[88] However, owner Roy Thomson, the Canadian press magnate, was determined on establishing the idea. Travel became essential to its coverage. The *Colour Section* of 7 April 1963 was a 'holiday boom' special. The cover showed the villa of the Dutch royal family in Italy and the issue devoted fourteen of its twenty-four editorial pages to travel in a forty-page issue. An introduction looked 'behind the allure of the brochures into the hard realities of the business'. One article reported, 'From its first Edwardian beginnings at Nice, the Riviera may soon stretch ... from Israel to Portugal. Greece, Morocco and other new-comers to the tourist field are now plunging heavily on new hotels, new resorts: for potential profits are vast.'[89]

Thomson's determination worked. By the autumn of 1963, issue sizes were up to fifty-two pages (20 October) and the five issues in March 1964 were all between forty-eight and sixty pages. Of twenty-two advertising pages in the forty-eight-page issue of 29 March, a third were travel-related. The links between the press and travel industries got even closer in 1965 when Thomson, by then a lord, bought up Universal Sky Tours, Britannia Airways and Riviera Holidays.[90] They became Thomson Travel, which in 1972 acquired Lunn Poly.[91]

Go magazine had two owners in the 1950s, and it was all change again in 1964 with a relaunch as 'the travel magazine for go-ahead people'. The first quarterly issue carried an article about fresh, familiar and faraway destinations. The fresh places were Portugal's Algarve, Spain's Costa D'Orada, Montenegro, Bulgaria's coast and 'Rumania's Riviera'. Jersey, Brittany, the Belgian coast, Majorca and Rimini Riviera were 'familiar'. The faraway places were Greece, Jordan, Morocco, Tunisia and Moscow.[92] Advertising showed British United Air Ferries flying from four airports.[93] The ups and

<hr>

88 Magazine timeline <http://www.magforum.com/time1.htm> accessed 6 October 2025.
89 Peter Wilsher, 'The expanding Riviera', *Sunday Times Colour Magazine* (7 April 1963), 12–17.
90 Anon., 'Press lord takes over Sky Tours, Riviera', *Travel Trade Gazette* (30 April 1965), 1.
91 Anon, 'Lunn Poly digs deep into history', *Travel Weekly*, (3 April 2000).
92 'Fresh, familiar and far-away', *Go* (1964, no. 1), 18–25.
93 Page advert. 'Fly with your car', *Go* (1964, no. 1), 45.

downs of the title show the problems of maintaining a travel magazine with sufficient sales.

Travel Supports New Magazine Sector

The 400-seat Jumbo jet landed in 1970. The era of mass aviation had begun. Over the next decade, more time off encouraged a second holiday and persuaded many workers to go away for a fortnight at a time. The English Tourist Board encouraged weekends away in Britain.[94] BBC TV launched the *Holiday* programme in 1969. In 1973, travel book publisher Lonely Planet was founded. Most magazine launches were in niches, or were crossover titles, such as Green Pea Publishing's *Food and Travel* in 1977. *Business Traveller* launched in 1976 and grew to become the leading magazine for frequent fliers with nine regional editions. The print unions went on to secure a 37.5-hour week in 1980 and five weeks of annual holiday in 1986.[95]

As Thomas Cook realized, magazines can be a powerful marketing tool. Airlines followed his example with in-flight magazines. British Airways' *High Life* was published by Headway, a pioneer of contract publishing in the 1980s. These publishers created magazines for the customers of large companies and offset the costs by selling advertising. Redwood was a leader in the sector from 1983, initially with travel titles: *Expression!* for American Express cardholders; *InterCity* for business train travellers; *Airport*, a pre-flight magazine at airports; and *Red Book*, a teletext booking system directory. Contract magazines often had large circulations – half a million in the case of *Expression!* – and *InterCity* became the best-read business magazine.[96] Redwood led the formation of the Association of

94 Smith, *The History of Tourism*, 6.
95 McKitterick, *A History of Cambridge University Press*, 361.
96 National Readership Survey, 1990.

Publishing Agencies, a trade body that evolved into the Content Marketing Association.[97]

Circulations of paid-for magazines were in decline. *Picture Post, Illustrated, Everybody's* and *John Bull* had a combined sale of 4.5 million copies a week in 1950, falling to 1.5 million in 1958.[98] The Sunday supplements finished them off. In 1955, *Radio Times* claimed the largest sale of any weekly – almost nine million. By 1983, it was still Britain's biggest-seller, though at three million copies. Holiday advertising was a big earner, for example the issue of 31 December 1983 included a sixteen-page holidays section. *Radio Times* lost its place as the top-seller to *Reader's Digest* in 1993.[99] In 2002, *Sky*, the satellite broadcaster's eponymous magazine, claimed the highest circulation at 5.2 million.[100] The *Thomas Cook Magazine* was produced by Redwood before it was taken back in-house in 2002. The contract sector also provided opportunities to experiment with the print format. In particular, Virgin Atlantic's *Hot Air* turned into the fanzine-influenced *Carlos* in 2005.[101]

In 1988, BBC Enterprises took control of Redwood and launched titles related to TV and radio content. One of the first was *Holiday '89*, the 'BBC good holiday guide'. This was not a success. However, Redwood revolutionized the food and gardening sectors with *Good Food* and *Gardeners' World*. *BBC Holidays* followed in February 1992 with a cover-mounted guide to Paris. However, the 'monthly magazine that takes you places' was closed three years later, despite a relaunch (Figure 7.3). Two factors in its

97 Anthony Quinn, 'Contract, custom and customer publishers', Magforum <http://www.magforum.com/custom_publishers.htm> accessed 6 October 2025.

98 Anthony Quinn, 'The slow death of the weekly magazine', Magforum <https://magforum.wordpress.com/2015/12/19/the-slow-death-of-the-weekly-magazine/> accessed 6 October 2025.

99 Anthony Quinn, 'Film, TV and radio magazines', Magforum <http://www.magforum.com/glossies/film_tv_radio_magazines.htm> accessed 6 October 2025.

100 'Magazine launches & events 2002', Magforum <http://www.magforum.com/2002.htm#joh> accessed 6 October 2025.

101 '£10 to New York and the inflight magazine', Magforum blog <https://magforum.wordpress.com/2015/03/17/10-to-new-york-and-the-inflight-magazine/> accessed 6 October 2025.

failure were that newsagents had no specific place to put it – there were no similar magazines with the sales to justify national distribution – and sales varied greatly across the year.

Figure 7.3: *BBC Holidays* (April 1995) was the lone travel magazine, so newsagents had no specific place to put it and sales varied greatly across the year (author's collection).

Magazines – 150 Years After Cook's First Trip

In 1991 there were 2,434 consumer and special interest publications and 4,608 publications classified as business and professional.[102] Of the consumer titles, seventy were listed as tourism-related. These were subdivided into:

- Tourism, 22, including: *Arab Traveller, In Britain* (contract title for the British Tourist Authority with a monthly circulation of 71,009), *Manx Post*, and *Konnichiwa*.
- Travel, 27: *Action Holiday* (50,000 print run), *Cruises* (through travel agents), *World Magazine* (monthly, 77,000 run).
- Inflight, 21: *Concorde Guide, Dream Journeys* (contract title for BA in Japanese), *High Life* (275,000 run).

Of the business and professional titles, 140 were tourism-related:

- Amusements and theme parks, 4.
- Hotels, 5.
- Railways, transport and navigation, 81.
- Travel, 50, including: *Business Traveller* (39,397); *TTG* (24,485); *Tours & Excursions*; *Travel Agency* (free monthly, 11,574), *Which Airline*.

In the 1990s, adventure was at the heart of tourism marketing – and magazine development. *Wanderlust* focused on adventurous travellers from 1993. A year later, the BBC turned to the publishing company Rough Guides for inspiration. As one TV review started: 'If the words travel show usually inspire you to nothing more than a trip to the fridge, try tuning in to *Rough Guide*, the BBC's appealingly eccentric, fast-paced twist on the genre.'[103] Publishers continued to attract travel advertisers, for example *Harpers & Queen* attached a forty-eight-page winter supplement to its October 1995 issue.

---

102   *BRAD (British Rates and Data)*, November 1991.
103   Jessica Shaw, 'Rough Guide on the BBC', *Entertainment Weekly* (19 August 1994) <https://ew.com/article/1994/08/19/rough-guide-bbc/> accessed 11 June 2022.

In 1997, *Condé Nast Traveller* rediscovered the reforming agenda of the British Guild of Travel Writers (Figure 7.4). The US edition, *Condé Nast Traveler*, had been founded in 1986 by former *Sunday Times* and *Times* editor Harry Evans. He set *Traveler* out as 'truth in travel'. The British

Figure 7.4: *Condé Nast Traveller* (October 1997): 'Why be a tourist when you can be a traveller?' (author's collection). The model Kate Moss was on the December cover.

version was launched under editor Sarah Miller. She told the *Guardian*: '[A press trip] is uniform and controlled by the PR company ... Travel isn't a controlled experience. It is what happens to you'. *Traveller* has a policy of not accepting free trips, to ensure editorial independence. It was able to avoid the fate of *BBC Holidays* for two main reasons. First, Condé Nast was able to explore the market by importing the US edition. Second, *Traveller* was not a mass-market magazine, but a luxury monthly, so it sat alongside *Vogue* on newsagents' shelves. The advertising slogan cemented the elite/mass-market language divide in a campaign with a budget of £1.2 million: 'Why be a tourist when you can be a traveller?'[104] Six images used in the campaign were colour paintings of distinctive destinations such as Venice. Miller said: 'We feel the use of illustration signals editorial in a far greater way [than photos]. This is a magazine, not a tourist brochure. The ["Why be a tourist"] tagline takes it further, giving you the context of the editorial.'

A year later, the *Sunday Times* assessed ten travel magazines and categorized them by reader: *National Geographic* (for armchair travellers); *Wexas Traveller* (armchair); *Wanderlust* (strapped for cash or armchair); *Family Travel* (paying for family); *Food and Travel* (broad range); *Traveler* (truly loaded, on expenses or armchair); *Traveller* (truly loaded, dirty weekend or armchair); *Harper's & Queen* (truly loaded, on expenses or armchair); *Ski and Board* (paying for family or dirty weekend); *What Cruise* (truly loaded or dirty weekend).[105] The next decade saw a proliferation of niche titles. Owners Abroad was a new type of holiday company, which marketed rental properties for their owners. Travel and property investment became intertwined. Merricks Media was founded in 1998 and published magazines with a lifestyle approach. These included *Florida Magazine*, *French Magazine*, *Greece Magazine*, *Italian Magazine* and *Spanish Homes*.

Other launches included crossover title *Blend* – style, music and travel – from 4130, an extreme sports publisher. *Spa Secrets* was a lifestyle quarterly. Style guru Tyler Brûlé launched *Line*, a *Wallpaper* spin-off covering sport, health, travel and fashion. Emap, one of the largest publishing groups, brought out *Escape Routes*. Contract publisher Absolute

104 'Publicis emphasizes *Traveller's* exclusivity', *Campaign* (12 September 1997), 7.
105 Joanna Hunter, 'Glossies jostling for prime position', *Sunday Times* (31 January 1998).

launched *Abta Magazine*, for the Association of British Travel Agents, and tried a consumer version, *Travelspirit*, in 2002. In 2003, the *Sunday Times* renewed its enthusiasm for magazines with *Travel*, produced on its behalf by River, a contract publisher. At the same time, the *Observer Travel* supplement helped improve the Sunday paper's circulation.[106] And celebrities saw the potential. *Jaunt* revealed the 'jet-set style secrets' of German model Heidi Klum. Her readers were 'looking for the golden age of travel and glamour'. *Wallpaper*'s publishers aimed for upmarket readers with *Navigator* in 2004. The launch press release claimed: 'Too many city guides [are] out of touch with the cosmopolitan thirty-something audience that make up the vast majority of people taking city breaks.'[107] Channel 4 television came out with a *Place in the Sun* spin-off. It became the best-selling overseas property title.

The effect of online growth on magazines became clear in 2006 when *TV Times* launched the Myholidayideas website to complement its classified travel pages. In 2007, the BBC's commercial arm took control of Lonely Planet, which sold 6.5 million guidebooks a year, and launched *Lonely Planet Magazine*. The move caused uproar. *Wanderlust* and *Time Out* accused the publicly funded BBC of distorting the marketplace.[108] The niches multiplied, as did the number of magazines with 'Traveller' in the title. The newcomers included *Football Traveller* (1998); *Culture Trip* (2001); *New Vistas* – 'property, travel and pure indulgence' (2005); *Lonely Planet Traveller* (2009); easyJet *Traveller* (2010); *Travel and Leisure* (2011); *Sidetracked* travel and extreme sports (2014); *Adventure Travel* (2016); *On Foot Traveller* (2016); *National Geographic Traveller* supplements (2018); *Silver Traveller* (2019); *Gourmet Traveller* (2019); and *Family Traveller* (2019).

Thomas Cook took 485 passengers on his first train journey. In 2018, the company took two million holidaymakers on its jets and ships. The travel giant was in a tough period, however, and was 'slowly going away'

---

106 *Media Week* (23 January 2003).

107 'Wallpaper group launches travel magazine', IPC press release (4 December 2003).

108 'Lonely Planet magazine in the shops', Magforum blog <https://magforum. wordpress.com/2008/11/27/lonely-planet-magazine-in-the-shops/> accessed 6 October 2025.

from brochures and moving towards 'magazine-style' publications. One of the reasons was fluid pricing, which meant brochures could be out of date before they were distributed.[109] It decided to revive the *Excursionist* name. However, it was not enough. Thomas Cook collapsed into administration. As for magazines, sales had been falling for decades. The Covid-19 outbreaks exacerbated this. In 2020, *Sunday Times Travel* closed. It had an average monthly circulation of 53,568. The ABC report listed only three consumer travel titles with circulations large enough to justify the cost of verification: *Condé Nast Traveller* (80,258), *Business Traveller* (67,263) and *National Geographic Traveller* (49,723). Alongside them were the trade journals *Travel Weekly* (13,343), *TTG* (10,533) and *Travel Bulletin* (5,182).[110]

Finally: 'Traveller' or 'Tourist'?

Subtle pecking orders exist in the English language – consider the words royal, regal and kingly – and 'traveller' has achieved a cachet not ascribed to 'tourist' or 'excursionist'. An 1854 letter to *The Times* commented on the contempt of railway officials for 'excursionists', and, in 1882: 'Like sheep [the tourist] seems to require for his holiday resort the presence of many similar to himself'.[111] Thomas Cook appears to have been attuned to the distinctions, changing the title of the *Excursionist* to *Traveller's Gazette* in 1903 – 'in deference to the higher social status of its readers'. The launch of *Condé Nast Traveller* crystallized a century-old debate. The history of magazines suggests 'traveller' trumps 'tourist' every time.

109   Amie Keeley and Ben Ireland, 'Thomas Cook to shift away from brochures to magazines', *Travel Weekly* (23 October 2017) <https://travelweekly.co.uk/artic-les/290425/thomas-cook-to-shift-away-from-brochures-to-magazines> accessed 6 October 2025.
110   ABC media audit, January to December 2020.
111   'The tourist season has at last run itself out', *Times* (7 October 1882), 9.

8 The Glories of Scotland in Picture and Song: Jumping on the Festival of Britain Bandwagon?

This chapter introduces a comparatively modest book of Scottish songs published in the mid-twentieth century by a Glasgow music publishing house, Mozart Allan. Such song repertoire was bread-and-butter to provincial publishers, and at first glance *The Glories of Scotland: in Picture and Song* has little to differentiate it from many earlier or contemporaneous publications.[1] The book coincided with the Festival of Britain in 1951, and various clues suggest that this was the impetus behind this particular publication, demonstrating a keen eye for a sales opportunity.

Music Publisher Mozart Allan

The rapid growth of Glasgow during the late nineteenth century is well-documented, with increasing industrialization and commerce drawing in workers at all levels of society. For the literate working and middle classes, a demand for live entertainment was paralleled by the need for printed music to sing and play at home. Scottish songs and fiddle tunes were enjoyed at all levels, and this offered an opportunity that various music publishers were keen to take up. Some opened offices in London, or branches in different Scottish cities, to extend their reach. However, two firms opened and remained solely in Glasgow, continuing to publish inexpensive new material until the middle of the twentieth century: James Speirs Kerr (1841–93), and Mozart Allan (1857–1929). While their output

---

1 Jack Fletcher (foreword), *The Glories of Scotland in Picture and Song* (Glasgow: Mozart Allan, 1950).

was by no means solely limited to these Scottish genres, their national repertoire survived the longest, and to judge by the number of copies available in libraries and in the second-hand market, must have been particularly popular. Mozart Allan's father was a dancing teacher for many years, and Mozart initially worked with his father, before branching out into publishing.[2] It is unclear how much control Mozart Allan Jun. had over the firm by the 1950s, since Jack Fletcher was proprietor of the firm by 1950, describing Mozart Jnr as a 'valued friend' in his Foreword to *The Glories of Scotland*. Fletcher was Chief Executive until at least the mid-1960s.

Scottish Music Editions: Outward Appearance

Books of Scottish music often had a distinctive and predictable outward appearance. Cloth covers were embossed, or paper covers printed, with thistles, Celtic knots, or the lion rampant of the Scottish Royal crest, but if one thing spelled 'Scottish' more than anything else, it was the use of tartan. Queen Victoria and Prince Albert's love affair with Balmoral, the Highlands and all things Scottish, contributed towards tartan becoming the symbol par excellence for Scotland. Indeed, the trend has disparagingly been called 'Balmorality'.

Tartan appeared on tourism promotional materials, gifts and souvenirs and postcards by manufacturers like Valentines, the Dundee postcard manufacturer who turned out hundreds of postcards over the years, along with calendars and other novelties. Starting in the late Victorian era, for at least half a century, tartan covers were a popular choice not only for miniature books of poems by Robert Burns, or Walter Scott's *The Lady of the Lake*, but also for Scottish songbooks, the publishers sometimes even offering spun silk tartan covers in the buyer's choice – ideal gifts either at home or

---

2   Always known as Mozart Allan, his full name was actually Ebenezer James Mozart Allan. The name can cause some confusion, since he and his brother also each named a son Mozart. Mozart Allan, the father, died in 1929, his unmarried son Mozart continuing to live in the family home with his maiden aunt, and then alone until he died in 1969.

for the Scottish diaspora.[3] Newspaper advertisements would particularly highlight them at Christmas and New Year.

In the 1950s, various Mozart Allan Scottish song collections were reprinted in differing iterations of diagonal red tartan. An early paperback published prior to 1902, *Two Hundred and Twenty Popular Scottish Songs (Tonic Sol-Fa Notation)* reappeared in a diagonal tartan reprint in the 1950s. Similarly, their very popular 240-page album, *The Morven Collection of Scottish Songs*, first published in 1890, was reprinted and advertised until the 1950s. It latterly appeared with either silk tartan, or diagonal tartan paper pasted onto the boards, with a red cloth spine and corners. It was advertised in Fraserburgh for 15*s.* in 1951.[4] Whether this was the silk or tartan paper version is unspecified. Mozart Allan's [Fifty] *Selected Songs of Burns*, first published in 1896, reappeared around the same time in gold-embossed diagonal tartan.[5]

New Titles by 1951: *Scotland Calling!* and *The Glories of Scotland*

Two new titles were on sale by 1951. Drawing on a catchphrase from the early days of radio, *Scotland Calling!* had been available in stiff boards for 5*s.*, or paper covers initially for 2*s.* 6*d.*, since *c.*October 1948, advertised in the *Oban Times and Argyllshire Advertiser*.[6] Again, it used a single diagonal tartan behind the title and contents blocks. Meanwhile, the paper-tartan-on-boards idea carried across to *The Glories of Scotland: in*

---

3   Spun silk was not the highest quality silk – which might explain why surviving copies have sometimes become a little shabby with age.

4   Findmypast Newspaper Archive Limited and British Library, 'British Newspaper Archive', 2020 <https://www.britishnewspaperarchive.co.uk/>, accessed 30 September 2021.

5   Robert Burns, *Selected Songs of Burns. Arranged with Symphonies and Accompaniments for the Pianoforte.* [Also Known as *50 Selected Songs of Burns*], Burns Centenary (Glasgow: Mozart Allan, 1896).

6   C. MacKay Collier, *Scotland Calling!: In 50 Scottish Songs, Staff, Sol-Fa and Words, for Community Singing* (Glasgow: Mozart Allan, 1948).

Picture and Song, now with a blue cloth spine and corners. Instead of a single diagonal tartan, fourteen different tartans appear in rectangles, the clan name underneath each one. The same arrangement appears on the reverse of the volume, with an advertisement for another book taking the place of the front title. It only appears to have been published in this hardcover format, at the price of 7s. 6d.[7] Both books were significantly shorter than *Morven*, at seventy-six and seventy-four pages respectively, and the piano accompaniments notably simpler. The new titles parted company from the earlier *Morven* and *Selected Songs of Burns*, in dispensing with a separate stave for the singer's line and instead laid the lyrics under the melody in the piano part. However, Fletcher continued to supplement the staff notation, that is standard music notation, of these new compilations with Tonic Sol-Fa notation above the singer's line, as in the earlier collections.

Focusing on *The Glories of Scotland*

Mozart Allan's publications are generally undated. Newspaper advertisements often provide clues, although they may not tell the whole story. *The Glories of Scotland* can be traced as first advertised in the *Oban Times and Argyllshire Advertiser* in October 1950.[8] Evidence from the Festival of Britain's Scottish Committee minutes corroborates that it would have been published by late 1950, as will be explained in due course. On the back endpaper of *The Glories of Scotland*, an advertisement for *Scotland Calling!* shows the cover of that collection, including amongst the contents, 'God save the King'. George VI died in February 1952.[9] In terms of an initial date, there is additional evidence that the photos could not have been taken

---

7 For comparison, 15s. would be £39 today. 7s. 6d. would be £19.50, 3s. would be £7.80, and 2s. 6d. would be £6.50. Kate Rose Morley, 'Historical UK Inflation Rates and Price Conversion Calculator' <https://iamkate.com/data/uk-inflation/>, accessed 24 June 2022.

8 *Oban Times and Argyllshire Advertiser* (28 October 1950), British Newspaper Archive, accessed 27 July 2022.

9 Only early printed copies will contain 'God save the King'. Subsequent copies contain 'God save the Queen'.

before 1949; the Glasgow illustrations show trolleybus wires supporting poles, which were installed in 1949.

The significant difference between the two collections is that only *The Glories of Scotland* features black and white photographs and a line-drawn map of Scotland.

The Glories of Scotland foreword is brief but breaks with convention and signifies the significance of paratext in song collections by naming such collaborators as the newly formed Scottish Tourist Board, founded in 1945; and British Railways, nationalized in 1948, and includes a reference to his 'valued friend Mozart Allan, whose untiring efforts are reflected in each and every page'.[10] Fletcher also obtained a photograph from the railway historian and photographer Ernest Richard Wethersett (1893–1987), famed for his writings about London steam trains and the LNER (London North Eastern Railway).[11] On the other hand, Aird & Coghill were the Glasgow printing firm responsible for many Scottish music publications; public recognition would be welcomed, but their input is not noteworthy in this respect.

Fletcher's commendation of the volume to 'all music lovers and true patriots [...] the world over', is unremarkable at first sight – for Scottish songs are enjoyed worldwide, particularly by emigrants – and yet it is still perhaps significant if we take it as an invitation for visitors to Scotland from all parts of the globe to take this uniquely Scottish product home with them.

An itemized 'List of Illustrations' follows the foreword, with twenty-two photographs of various urban and rural Scottish locations scattered throughout the book. Illustrations do occasionally appear in Scottish songbooks, but generally in connection with the poet Robert Burns, or locations forming the subject of his poems, so the scope of illustrations in *The Glories of Scotland* is broader than usual. A photograph of a contemporary or famous earlier proponent of Scottish song is also often to be found either inside or printed on a paper cover. In this case, in addition to the

---

10 British Railways traded as British Rail from 1965 until privatization in the late 1990s.
11 E. R. Wethersett and others, *Locomotives of the LNER: A Pictorial Record* (1st ed.) (Cambridge, London: W Heffer and Sons, 1947).

Scottish organizations and individuals named in the foreword, the opposite page bears a 'signed' photograph of Alec Finlay, captioned 'Scotland's own comedian'. Finlay (1905–84), who performed a character known as 'Scotland's Gentleman' in variety theatre first in Scotland and then in London, gave Fletcher permission to reproduce the text of one of his recent songs, co-written with the Greenock comedy script-writer Bill McDonnell (*c*.1906–54). In August 1950, the *Port-Glasgow Express* had reported *Let Scotland Flourish* to be a best seller in America and Canada, very popular with clan societies, and observed that it augured well for Finlay's visit to America with singer Robert Wilson that autumn. The words for *Let Scotland Flourish* appear in *The Glories of Scotland*, but, tellingly, a footnote advises the reader that any music seller could supply the 'Complete Words, Music & Solfa' – this single song sheet was also a Mozart Allan publication.[12] Endorsement by a showbusiness celebrity would have made the book more saleable, although buyers might have been disappointed to discover that the music for the song was not in fact contained in it.

Finally, on the verso of Finlay's photograph, is a map, followed by the contents list. The map is noteworthy, for this is not a usual inclusion in a songbook. Major towns and cities, lochs and significant tourist sites such as Balmoral, Braemar, Burns Cottage and Monument near Ayr and Walter Scott's Abbotsford are indicated, some with thumbnail illustrations, along with an aeroplane symbol at Prestwick, which was the international airport for the West of Scotland at that time – Pan American Airways had begun flights to Glasgow Prestwick in 1946. No roads or railway lines are indicated.

The list of illustrations indicates the Scottish region where these beauty spots are located, ranging from the Scottish Borders; Ayrshire; the Central Belt (with Glasgow on the west and Edinburgh on the east coast); Argyll and Bute; the Trossachs; Perthshire and Dundee; and as far north as Inverness. The attributions to contributors are not indicated in

---

12　Macdonnell, Bill and Alec Finlay, *Let Scotland Flourish; Written and Composed by Bill McDonnell & Alec Finlay* (Glasgow: Mozart Allan, 1950). The poem on p.71 of *The Glories of Scotland* appears 'By kind permission of Alec Finlay'. The song was recorded and issued by the Canadian label, Scottish Clan Records, in 1954, sung by Finlay on the A side and narrated on the B side.

the illustrations list, only appearing beneath the photographs themselves in the music part of the book. The Scottish Tourist Board was the major contributor. Moreover, although the list of illustrations gives no indication of the relationships between songs and photographs, Fletcher and/or Mozart Allan Jnr, in collaboration with their contributors, evidently attempted to link individual songs with photographs of their locality, as demonstrated in Table 8.1.

Post-War Tourism

When the Second World War ended amidst the wave of optimism and determination to rebuild every aspect of the United Kingdom's economy, tourism was recognized as an important sector, and Scottish hoteliers and travel firms were amply aware of the country's attractions. Although the official Scottish Tourist Board was not founded until the 1969 Development of Tourism Act, a demand for a Scottish Tourist Board was evident by October 1945 and by the end of the year a chairman had been announced:

> A new Scottish tourist board under the chairmanship of Mr Tom Johnston, former Secretary for Scotland, has been appointed by the Scottish Council on Industry. Sir Douglas Thomson. M.P., and Mr Robert Wotherspoon, Inverness, chairman and vice-chairman of the Scottish Travel Association, will serve on the new board. It is hoped to arrange for the merging of the activities of both organisations.[13]

Bus day trips and extended touring holidays, steamer rides on Loch Katrin and Loch Lomond, 'all-in' package holidays and self-drive car rentals were already some of the attractions on offer by 1950, as seen in the Scottish Tourist Board's magazine, *Take Note!* which was aimed at would-be visitors to Scotland and clearly anticipating a wave of visitors with clan connections from the diaspora. The magazine offered a variety of articles about Scottish people, places and history, with advice on where to stay, hotel

---

13   *Aberdeen Evening Express* (Thursday 27 December 1945), British Newspaper Archive, accessed 20 July 2022.

advertisements and several pages of diary listings for notable events around Scotland. Similarly, a British Railways leaflet, 'Through the Trossachs', published in 1950, offered a circular tour of the Trossachs, with various stretches by 'train, steamer and motor', and a wide selection of different tickets including cheap day, evening or excursion tickets, Clyde Coast season tickets, or combined day tours. The proximity of the Trossachs – the southernmost part of the Scottish Highlands – and Lochs Lomond and Katrine to Glasgow made them appealing tourist destinations.

While very low compared to today, car ownership in Britain was rising.[14] Increased mobility and increasing leisure-time made travel more appealing to British citizens as much as to foreign visitors. Moreover, an extra boost to tourism would have been given by the announcement that finally, after several years of fluctuating arrangements, petrol rationing was abolished altogether on 26 May 1950 – just in time for the late May Bank Holiday and welcome news to all drivers.

The Festival of Britain, 1951

The tourist trade was already a significant part of Scotland's economy before the Festival of Britain moved everything up a gear. Although initially envisaged as a giant international trade fair, the Festival of Britain eventually became a national celebration of Britain's achievements in many spheres, intended to act both as a 'tonic to the nation' (an expression originating in a letter from Gerald Barry, editor of the *News Chronicle*, to Director of Science, Ian Cox) and to attract international visitors, as well as marking the centenary of the Crystal Palace Exhibition.[15] Sited on London's

14   J. F. Kain and M. E. Beesley cite J. C. Tanner's calculations of 0.16 per capita car ownership between 1945 and 1951, 0.22 between 1951 and 1956, and 0.37 between 1956 and 1960 'Forecasting Car Ownership and Use', Urban Studies, 2.2 (1965), 185.

15   F. M. Leventhal, '"A Tonic to the Nation": The Festival of Britain, 1951', *Albion: A Quarterly Journal Concerned with British Studies*, 27.3 (1951), 445–53.

South Bank, there were also exhibitions in Glasgow at the Industrial Power Exhibition at Kelvin Hall; Edinburgh at the Living Traditions of Scotland, in the Royal Scottish Museum; and the Farm and Factory exhibition in Belfast. Furthermore, a travelling road exhibition and a floating exhibition on a repurposed naval ship, the Festival Ship Campania, was in Scotland between 17 September and 6 October 1951. There were other exhibitions such as the twentieth-century and eighteenth-century Scottish book exhibitions in Glasgow's Mitchell Library and Edinburgh's Signet Library respectively; and a multitude of nationwide concerts, marches, competitions, parks and heritage building improvements.[16] The organization of this year-long national celebration obviously commenced well before 1951; indeed, the first meeting of the Festival of Britain's Scottish Committee was in June 1948. The Festival was intended to appeal to the broadest possible audience both at home and abroad and, with overseas visitors in mind, publicity began a full year earlier in Australia and New Zealand.[17] A contemporary British Railways advertisement in North America exemplifies both the tone of the advertising and the scope of the Festival:

ALL BRITAIN IS THE SCENE OF THE 1951 FESTIVAL OF BRITAIN!

> Imagine an entire nation on parade! That's Britain in 1951, with EXHIBITIONS, PAGEANTS, SPORTS EVENTS, FESTIVALS of MUSIC and DRAMA *everywhere* in England, Scotland, Wales and Northern Ireland.
>
> Starting from London, centre of this gala programme, you'll want to visit many of the cities and towns playing their specialized roles in this great Festival.
>
> In Britain, *travel* means BRITISH RAILWAYS. By securing **all** your travel needs before you leave home, you will realize substantial savings NOT obtainable in Britain ... for example MILEAGE COUPONS, which permit *go-as-you-please* travel. [...] Leave with assured reservations on trains, on cross-channel steamer services [...] and at any of the 47 outstanding hotels operated by The Hotels

16   Mary Banham and others, *A Tonic to the Nation: The Festival of Britain 1951* (London: Thames and Hudson, 1976).

17   Gerald Barry, 'The Festival of Britain: 1951', *Journal of the Royal Society of Arts*, 100.4880 (1952), 667–704.

Executive, British Transport. Arrangements for sightseeing trips and tours by rail, motor coach and steamer, can also be completed *before* you go abroad.

PLEASE CONSULT YOUR TRAVEL AGENT or any British Railways Office shown below: [addresses in New York, Chicago, Los Angeles and Toronto].[18]

Another contemporary British Railways advertisement specifically highlighted the charms of the Scottish scenery:

$376 to see THE TROSSACHS
Scotland's Scenic Wonderland!
(From Edinburgh or Glasgow)
Travel in Britain is amazingly inexpensive!
So, stay longer and see more. Make sure of maximum savings – buy ALL your British transportation BEFORE YOU LEAVE.[19]

A little closer to home, four double-decker buses were despatched round Europe in autumn 1950, their top decks fitted out as miniature exhibitions to publicize the Festival.[20] The Festival of Britain Office produced a series of thirteen *About Britain* tourist guides published by William Collins Sons & Co., of 14 St James's Place, London, each with aerial photographs, an introductory essay by a well-known author, road maps, drawings and notes, all for 'a reasonable price' of 3*s. 6d.*[21] Books 11 and 12 were devoted to the Lowlands of Scotland and the Highlands and Islands respectively.[22]

18 Advertisement being sold on eBay, accessed 17 July 2022, source not given.
19 Another advertisement being sold on eBay, by the same supplier and possibly from the same unspecified magazine.
20 Meanwhile in Scotland, local authorities were clearly expected to subsidize the Scottish Tourist Board; Glasgow Council Minutes record a request from the Scottish Tourist Board for more financial backing towards the end of 1950, suggesting that advertising was perhaps going over-budget.
21 £6.40 in 2022.
22 Geoffrey Grigson and others, *About Britain*, nos. 1–13 (London: Collins for the Festival of Britain Office, 1951); John Robertson Allan, H. G. Stokes, and Festival of Britain Office, *Lowlands of Scotland, About Britain*, 11 (London: Published for the Festival of Britain Office by Collins, 1951); Alastair Dunnett, *Highlands and Islands*

Despite letters to local newspapers objecting that the money could be spent on more worthy causes, it appears that the British public mostly embraced the Festival of Britain, to judge by the sheer number and scope of events put on between May and September 1951. Barry wrote that although the festival was primarily organized 'for the people of Britain themselves', attracting visitors from abroad was obviously 'one of its major aims'; and in fact, visitors to Britain that year rose from 608,000 in 1950, to 700,000 in 1951, exceeding expectations by tourist organizations.[23]

Newspaper articles about Glasgow International Festival events do not mention the Mozart Allan publishing house; and the few newspaper advertisements for *The Glories of Scotland* – for example, listed amongst various Scottish music publications sold by C. Dennis in Fraserburgh, thirty-seven miles north of Aberdeen – do not mention the International Festival.

However, the paucity of newspaper advertisements does not necessarily argue against the thesis that Fletcher's compilation was intended for tourist consumption. International tourists might arguably not buy many provincial newspapers during their stay, particularly if they had booked organized tours which whisked them between locations. I suggest that advertising *The Glories of Scotland* solely in late 1950 and 1951, strongly suggests a tourist market having been envisaged. There was the expectation that there would be more visitors to the United Kingdom that year and that Scottish tourism would benefit. Although Glasgow's Exhibition of Industrial Power at Kelvin Hall between 28 May and 18 August attracted fewer visitors than expected, tourism to Britain rose significantly that year which suggests that Scotland attracted its fair share of tourists.[24]

---

of Scotland, *About Britain*, 12 (London: Published for the Festival of Britain Office by Collins, 1951).

23 Barry, 'The Festival of Britain: 1951'.

24 William Hepburn, 'Glasgow's Forgotten Exhibition: The Festival of Britain at Kelvin Hall, 1951', *Welcome to the Kelvin Hall Project Blog*, 2016 <https://kelvinhallproject. wordpress.com/2016/04/07/glasgows-forgotten-exhibition-the-festival-of-britain-at-the-kelvin-hall-1951/> accessed 15 July 2022; see also Michael Gallagher and Glasgow City Archives, '"Spectacular Story": Memories of Glasgow's Industrial Power

Moreover, Glasgow's Mitchell Library hosted the 'Exhibition of 20th Century Scottish Books'. The Scottish Committee of the Festival of Britain delegated the work of organizing both the Glasgow and Edinburgh exhibitions to a subcommittee chaired by William Beattie, Keeper of Printed Books at the National Library of Scotland. The organization of the exhibitions was in turn delegated to Perth City Librarian R. O. Dougan, granted a year's leave of absence in which to do the work.[25] In July 1950, Scottish publishers, printers and bookbinders were invited to contribute their publications, and about 2,000 post-war books were selected, along with another 3,000 pre-war titles mainly sourced from the Mitchell Library's Collections. In early December 1950, the Scottish Books Exhibition Committee reported that there had been a good response, and books were being catalogued, so that by 16 February 1951 the catalogue was scheduled to be sent to the printers. The exhibition catalogue had a section for Scottish Music and Drama extending to four pages listing music and a further two pages of 'Books on music and drama by Scottish authors'.[26] Six Mozart Allan publications are listed among the music items, including the two new songbooks, *Scotland Calling* and *The Glories of Scotland*.[27] Jack

---

Exhibition in 1951', *Glasgow Times* (5 June 2022) <https://www.glasgowtimes.co.uk/news/20177775.amp/> accessed 15 July 2022.

25 'His Task Among The Tomes' *Dundee Evening Telegraph* (Dundee, 16 August 1950), *British Newspaper Archive*, accessed 20 July 2022.

26 The commercially published bibliography, Festival of Britain. Scottish Committee, Scotland) Mitchell Library (Glasgow, and Robert O. Dougan, *Catalogue of an Exhibition of 20th Century Scottish Books at the Mitchell Library, Glasgow [1 June to 28 July 1951]* (Glasgow: Scottish Committee of the Festival of Britain, 1951) is available at several libraries; a presentation copy documenting the organization of the exhibition and containing both the Glasgow and Edinburgh bibliographies can be consulted at the Mitchell Library itself: R. O. Dougan, Mitchell Library (Glasgow, and Festival of Britain, *Exhibitions of Scottish Books at the Mitchell Library, Glasgow and the Signet Library, Edinburgh/Organised on Behalf of the Scottish Committee of the Festival of Britain 1951 by R.O. Dougan*, 1951.

27 *Allan's Collection of Reels & Strathspeys, Quadrilles, Waltzes, Country Dances, Highland Schottisches, Jigs, Hornpipes &c &c. Arranged for Pianoforte* (Glasgow: Mozart Allan, 1883); Robert Burns and Arthur De Lulli, *Selected Songs of Burns. Arranged with Symphonies and Accompaniments for the Piano Forte [Cover Title:*

Fletcher responded to the invitation by submitting those six titles – four 'heritage' and two new. It can be assumed that he offered the older titles which he felt were either the most representative or the most likely to be of interest, along with two new titles, the latest of which was published with the International Festival in mind. As noted earlier, *The Glories of Scotland* was published by the autumn of 1950, in time for the catalogue in preparation for early December 1950 and printed in mid-February 1951.

Fletcher sourced his images primarily from the two organizations with the most invested in the success of the Festival of Britain for Scotland: the Scottish Tourist Board and British Railways. One of the photographs in *The Glories of Scotland*, 'In the Trossachs' was also used in the British Railways leaflet, 'Through the Trossachs'. The majority of pictures in the songbook related to a geographical location named in an adjacent song.

What's in a Name?

'Scotland Calling' was an established catchphrase from the early days of broadcasting: Glasgow's 5SC was the fifth BBC radio station, and the first Scottish, to open in 1923 and when the Empire Exhibition was opened in Glasgow's Bellahouston Park in 1938, it was used as an advertising heading. By contrast, the 'Glory' or 'Glories' of Scotland' expression had been in circulation for much longer. Even if Fletcher had been unaware of an early piano selection, Maurice Cobham's *The Glory of Scotland*, published in London in 1859, he would probably have known of the popular Scottish journalist, John Joy Bell's illustrated travel guide, *The Glory of Scotland*,

50 *Selected Songs of Burns]* (Glasgow: Mozart Allan, 1948); Andrew. MacKenzie, 51 *Beauties of Scottish Song Adapted for Medium Voices with Tonic Sol-Fa* (Glasgow: Mozart Allan, 1921); Marr and Company., Marr & Co's Royal *Collection of Highland Airs, Quicksteps, Strathspeys, Reels & Country Dances. For Piano with Piano Accordion Markings*, Reprint of Marr & Co's original edition (Glasgow: Mozart Allan, n.d); Fletcher (foreword), *The Glories of Scotland; Collier, Scotland Calling!*

first published in 1932. Bell's book, which came with maps of northern
and southern Scotland at the back, had been written 'primarily for readers
who have never been in Scotland' and was described as 'a holiday book [...]
for the convenience of the traveller by rail and the traveller by road'.[28] The
title page bears a quotation from Dorothy Wordsworth's 1803 Journal,
'Coleridge hailed us with a shout of triumph ... exulting in the glory of
Scotland.' It was, indeed, an old expression. In other words, the music
publisher Mozart Allan's two new publications, *Scotland Calling!* and *The
Glories of Scotland*, would immediately have struck a chord with contemporary purchasers.

Categorizing Songs in *The Glories of Scotland* and *Scotland Calling!*

It is noteworthy that twenty-seven of the song titles in *The Glories of
Scotland* were also in *Scotland Calling!* The foreword of the latter states
C. McKay Collier was the musical arranger, but his name is not mentioned
in *The Glories of* Scotland.[29] Thus, it is not known whether the four titles
using different musical arrangements in *The Glories of Scotland*, were by
Collier or by someone else. A fifth song title had a choice of two different
settings in *Scotland Calling!* but only one setting made it into *The Glories
of Scotland*. Compared to this, nineteen songs were unique to *The Glories
of Scotland*, set by an unknown arranger (perhaps Fletcher or Mozart Allan
Jnr, if not Collier), while twenty-three were unique to *Scotland Calling!*
In terms of prominent poets, it is hardly surprising that the two collections duplicate six songs by Robert Burns, three of which are in different
settings; one by James Hogg; one by Lady Carolina Nairne; and three

---

28 John J. Bell, *The Glory of Scotland* (London: Harrap, 1932). The maps are unillustrated,
    but do indicate major railway lines. As noted above, Fletcher's map in *The Glories of
    Scotland* differs in both respects.
29 Collier was an organist and was a graduate of the Royal Scottish Academy of Music.

songs by Sir Walter Scott. Each also has songs by Burns and Nairne that are not duplicated across the two publications. They do not draw upon the arrangements in the much earlier Mozart Allan publication of [Fifty] *Selected Songs of Burns*, which was still in print.

If any attempt is to be made to categorize the kind of material typifying each collection, then it appears that *The Glories of Scotland* inclines towards songs that can be located geographically and perhaps has fewer recently composed songs. If the book was intended to appeal to tourists and travellers, then it would be logical to include songs that invoke a sense of place, and songs that were more 'traditional' than the music hall or variety theatre style. By comparison, the foreword of *Scotland Calling!* boasts that it contains 'several songs which have never before been published in a Volume of Scottish Songs'. Identifying new material at a distance of seventy years can only be tentative; the collection contains five songs which are definitely not 'traditional' and might have had a particular appeal for the émigré, but this does not mean that they were new: 'Hurrah for the Highlands!' had appeared in an earlier non-Mozart Allan publication; 'O sing to me the auld Scotch sangs' was again not new to this publication; and 'We're no' awa' tae bide awa' was played by pipers as naval ships left the quay during the First World War.[30] This leaves 'My heart warms to the tartan'; and 'Soft Lowland tongue of the Borders', both of which had appeared in the earlier *Morvern* collection.

The piano arrangements in the earlier book of Burns songs had not been difficult, but those in *The Glories of Scotland* and *Scotland Calling!* were even simpler. *Scotland Calling!* claims to suit the 'player of average ability' and the cover strapline suggests that the collection is 'For Community Singing'. Neither of these claims is made for *The Glories of Scotland*. However, given the overlap between the two books, it is not surprising that both have easy accompaniments, widening their appeal to amateurs.

---

30 'Alan G. Brydon, "Calling Doon the Line"' <https://war-poetry.livejournal.com/740789.html>, accessed 20 July 2022.

For International Visitors

Although *The Glories of Scotland* continued to be advertised on the back cover of *Scotland Calling!* and on contemporary copies of *Selected Songs of Burns*, there are very few copies of *The Glories of Scotland* in Jisc Library Hub Discover or WorldCat union catalogues of library holdings and comparatively few of *Scotland Calling!* (There are more copies overseas than in Britain in each case, but WorldCat does not reflect all public library catalogues, so not too much can be made of this.) On the second-hand market, there are more copies of *Scotland Calling!* on eBay, than of *The Glories of Scotland*. One can perhaps infer that fewer copies of the latter were printed.

All the evidence points to *The Glories of Scotland* having been produced with an eye to the tourist market during the *Festival of Britain* year. Before even contemplating the music, the unique arrangement of clan tartans on the cover is notable, compared to the single diagonal tartan on other contemporary Mozart Allan publications; the fact that it was a more expensive, hardback copy compared to the cheaper hardback or even cheaper paperback editions of *Scotland Calling!*; the use of photographs from the Scottish Tourist Board, British Railways and Mr Wethersett; not to mention the canny choice of titles for both collections and their submission for the 'Exhibition of 20th Century Scottish Books'. Meanwhile, the music appears to have been deliberately chosen for different markets: *The Glories of Scotland* for international visitors and *Scotland Calling!* for the domestic market and expatriates overseas.

If these hypotheses are correct, then they support the argument that even a seemingly humble and musically unremarkable book of songs can, if placed in the context of a particular time and place, reveal much more about the publisher's motivations than first impressions might suggest. From a tartan-covered collection of Scottish songs by a provincial publisher, we are led to a series of careful decisions aimed at selling a uniquely Scottish product to a specific audience in one memorable year: not exactly a musical tour, but an effort to align the tourists' travels with at least some songs linked to those locations.

Table 8.1: Photographs opposite songs in *The Glories of Scotland*

PICTURE	SONG OPPOSITE	SOURCE
George Square, Glasgow	Title page	Scottish Tourist Board
Scotland's Own Comedian, Alec Finlay	Foreword and List of Illustrations	Alec Finlay
Map of Scotland		
Trossachs – Ben A'an from Dukes Road. [southern Highlands, an early tourist destination near Glasgow]	The Blue bells of Scotland (about a Highland laddie)	British Railways
Loch Lomond at Tarbet, [Argyll and Bute]	The Bonnie banks o' Loch Lomond	Scottish Tourist Board
Galashiels Common Riding – Fording the Tweed at Abbotsford	Blue Bonnets over the Border; words by Walter Scott, whose home was at Abbotsford.	Scottish Tourist Board
The Burns Monument, Alloway, Ayrshire	Green grow the rashes, O!, words by Robert Burns	Scottish Tourist Board
Dryburgh Abbey, Melrose, Roxburghshire	The flowers o' the forest; poet Alison Cockburn was born at Fairnilee House, near Melrose.	Scottish Tourist Board
Holyrood Palace – side gates, Edinburgh	The Queen's Maries; set in Edinburgh, where Mary Hamilton was hung. The next song is Within a mile of Edinburgh town.	Scottish Tourist Board
Forth Bridge at South Queensferry – Nr. Edinburgh	Caller herrin'; linking to words 'new drawn from the Forth'	Scottish Tourist Board
Provand's Lordship & Cathedral – Glasgow	The auld hoose. The picture is unrelated. Unlike the poem's demolished 'auld hoose' with 'wee' rooms, the Provand's Lordship still stands opposite Glasgow Cathedral.	Scottish Tourist Board

(Continued)

Table 8.1: Photographs opposite songs in *The Glories of Scotland* (Continued)

PICTURE	SONG OPPOSITE	SOURCE
Dunoon & Firth of Clyde – Argyll	Mary of Argyle	Scottish Tourist Board
Burns' Cottage – Ayr. (1759–96)	The cottage whaur Burns was born [sic]	Same photo as on front of [50] *Selected Songs of Burns*
Loch Ness at Drumnadrochit – Inverness	Sound the pibroch; the song concerns highland glens and clans, referencing Loch Shiel on the west coast, and Culloden on east. Loch Ness runs much of the way between them.	Scottish Tourist Board
Royal Border Bridge – Berwick [train on the bridge]	Kirkconnel Lea; is indeed in the Borders; but Berwick is on the east coast while Kirkconnel is more to the west.	R E Wethersett
University & River Kelvin – Glasgow	Kelvingrove	Scottish Tourist Board
Edinburgh Castle	The hundred pipers; picture unrelated. Jacobite song, marching to the Borders for Bonnie Prince Charlie.	Scottish Tourist Board
Broomielaw – River Clyde, Glasgow	The Sons of Scotland. Ends, 'Who was it famous made the Clyde but the sons o' bonnie Scotland.'	Scottish Tourist Board
Tay Bridge – from Wormit Station [Dundee] showing the piles of the original bridge blown down in 1879.	Bonnie Dundee, followed by The piper o' Dundee.	British Railways
Valley of the River Tummel – Perthshire	The standard on the Braes o' Mar. Picture unrelated; Braemar is in Aberdeenshire.	Scottish Tourist Board

Table 8.1: (Continued)

PICTURE	SONG OPPOSITE	SOURCE
Princes Street – Edinburgh	*Let Scotland flourish*; the words of the song made famous by Alec Finlay.	Scottish Tourist Board
Loch Earn and Village of Lochearnhead – Perthshire	A guid new year; picture unrelated.	Scottish Tourist Board
In the Trossachs	Auld lang syne; picture unrelated.	British Railways (The photo also appears in a British Railways leaflet from 1950, 'Through the Trossachs')
Ben Nevis – from Inverlochy, Fort William, Inverness	Text. 'From scenes like these old Scotia's grandeur springs, that makes her loved at home, revered abroad.' (Lines from Burns' poem, 'The Cotter's Saturday Night'.) The photograph depicts majestic scenery but is unrelated to Saturday night in a humble farmworker's cottage.	Scottish Tourist Board

SARAH BURKE CAHALAN AND KAYLA HARRIS

9 Pilgrim Pamphlets: Pocket Guides to Christian Tourism in Twentieth-Century Britain

Shrine pamphlets, aimed at pilgrims, tourists, those travelling only spiritually by means of reading the written word, and everyone in between, represent a unique genre of print material. From site-specific prayers to listings of nearby restaurants, hotels, souvenir shops and attractions, these pocket guides illustrate the complex space that Christian shrines occupy for both devotion and tourism, therefore the printed material is multifunctional. The 1955 advertisement depicted in Figure 9.1 is printed on the back of a small guidebook titled *Walsingham: Everybody's Guide Book with Plans and Illustrations*. It can then be assumed that this particular guide is aimed at someone physically already at the site, with the very specific direction of 'opposite the village pump'. Notably, the shrine office is listed last, while the shrine shop selling statutes, books, medals, pilgrim tokens and all types of souvenirs, gets top billing on the page:

> The Shrine Office (opposite the village pump) deals with all business matters, the booking of organized pilgrimages, enrolling in the Society, propaganda literature, intercessions to be offered in the Shrine. Orders for making vestments and church linen, painting, gilding and artistic work of all kinds received.

Religious shrines and festivals attract millions of visitors each year, ranging from pilgrims on spiritual journeys to casual tourists, depending on a traveller's interior beliefs and motivations. The United Nations World Tourism Organization (UNWTO) states that 'tourism is a social, cultural and economic phenomenon which entails the movement of people to countries or places outside their usual environment for personal or business/professional purposes. These people are called visitors (which may be either tourists or excursionists; residents or non-residents) and tourism has to do

Figure 9.1: Advertisement.

with their activities, some of which involve tourism expenditure.'[1] Within this definition, religious pilgrimage would fall within the categorization of tourism although there is a spectrum of beliefs about how to distinguish between tourists and pilgrims. Pilgrimage is also a form of intangible cultural heritage, as defined by the United Nations Education, Scientific, and Cultural Organization (UNESCO), with multiple pilgrimages (across faiths) being included on the representative lists that are evaluated by that agency each year.[2] Pilgrimage as a form of heritage involves the destined site as well as the routes, liturgies, processions, prayers, fasting and/or special foods and other behaviours associated with peoples' interactions with the site. The pamphlets discussed in this chapter bring participants into the community of these practices.

Despite some differences between the needs of pilgrims and tourists, there are also many commonalities. Both must prepare for their travel by selecting their destinations and modes of transportation. Once at their destination, they learn about the shrine's history through its architecture, artwork and other notable qualities such as miraculous events or occurrences. They may participate in tours, special events, or rituals unique to that location and culture. They require lodgings and food while they are there. Finally, as they leave, they often take a souvenir, which can hold memories of the journey and transport them back. All of these steps often involve printed material, from guidebooks and maps for planning, to postcards and posters taken home as souvenirs. This chapter considers the print culture surrounding pilgrimage to British shrines to the Virgin Mary, both Anglican and Catholic, during World War II and in the two decades following the end of the war.

---

1 'Glossary of Tourism Terms', United Nations World Tourism Organization, <https://www.unwto.org/glossary-tourism-terms>, accessed August 1, 2022.
2 'Browse the Lists of Intangible Cultural Heritage and the Register of good safeguarding practices', United Nations Education, Scientific, and Cultural Organization, <https://ich.unesco.org/en/lists>, accessed November 7, 2022.

Common Ground with the Virgin Mary

The term 'Anglo-Catholicism' was first used in the early nineteenth century and it is connected primarily with the Oxford Movement and exponents such as Saint John Henry Newman (1801–90); however, it reflects centuries of both theological and popular initiatives within Anglicanism to interrogate the place within that tradition of such 'Catholic' traditions as the sacraments, liturgy, church architecture and aesthetics, and devotion to the saints.[3] While some members of the Oxford Movement (such as Newman) eventually found a home in Roman Catholicism, many remained in the Church of England – thereby influencing the practices of that church and all those international churches which are in communion with it. Devotion to the Virgin Mary is one area where many Anglicans and Catholics have found common ground, as documented by the numerous publications promoting and facilitating that devotion.

The pilgrimage site to Our Lady of Walsingham is one site where both Anglo-Catholics and Roman Catholics have demonstrated their enthusiasm for pilgrimage. The site was destroyed during the dissolution of the monasteries but revived in the twentieth century, benefiting from the confluence of Anglo-Catholicism, the lingering medievalism of the Gothic Revival, and the increase in tourism in Europe. In a sense, it also benefitted from the lack of historical documentation around its medieval heyday. As Simon Coleman writes, 'the fact that so little is actually known about medieval Walsingham provides a historical and quasi-mythical blank space in which multiple readings of the past can be inscribed.'[4] According to Coleman, this blank space allowed, and allows, Christians from multiple sects to feel at home at Walsingham, to coexist in the space, and even to engage in nostalgia for a long-ago period of Christian unity. Walsingham has become such a destination that not all visitors have pilgrimage as their

---

3   The Oxford Movement is a high church renewal movement within the Church of England, many members of which subsequently converted to Roman Catholicism.

4   Simon Coleman, 'Pilgrimage to "England's Nazareth": Landscapes of Myth and Memory at Walsingham', in *Intersecting Journeys: The Anthropology of Pilgrimage and Tourism*, Ellen Badone and Sharon R. Roseman, eds., (Urbana and Chicago: University of Illinois Press, 2004), 60.

primary goal; as with major cathedrals, people also come for the historical and cultural interest; it is a 'forum for exploration – of the medieval past, of group dynamics among one's fellow travellers, or new liturgies, of novel forms of behaviour – whether that means going to the pub or praying to the Virgin'.[5]

Walsingham has a rich paper trail of books, periodicals, pamphlets and realia/ephemera associated with pilgrimage. It is particularly popular due to its ecumenical appeal and will be the primary example discussed in this chapter. But it is far from the only Marian pilgrimage destination in Britain; others include Our Lady of Consolation in West Grinstead, Our Lady of Westminster, Our Lady of Willesden in London, The Church of Our Lady St Mary of Glastonbury, Our Lady's Shrine at Evesham, The Shrine of Our Lady of Jesmond near Newcastle, Our Lady of Guisborough in North Yorkshire, Our Lady of the Undercroft Chapel in Canterbury, Our Lady of the Taper (or Our Lady of Cardigan) in Wales, Our Lady of Haddington in Scotland, and more. Like Walsingham, many of these sites were dismantled under Henry VIII and restored in the twentieth century.

The authors of this chapter write from the Marian Library, a collection at the University of Dayton in Ohio, United States. The Marian Library was founded in 1943 by the Society of Mary, a Catholic religious order which also founded the university in 1850. The library, intended to be ready in time to mark the centennial of the university itself, started as a modest selection of books with minimal-to-none acquisitions budget. However, in the intervening decades it has become a significant inter-national resource for studying the Virgin Mary. In addition to books, prints, ephemeral materials such as stamps and holy cards, and art, the library holds numerous pamphlets and brochures documenting a variety of devotional practices – and significantly, often multiple editions of the same publication, making it possible to trace changes over time. Especially during the Marian Library's initial years after establishment, much of the material represents a particular viewpoint. Operated by Society of Mary, Marian Library employees would write letters to their networks or to other libraries in the United States and in Europe, asking for donations

---

5 Coleman, 'Pilgrimage to "England's Nazareth": Landscapes of Myth and Memory at Walsingham', 62.

of material. Later, employees would specifically go on acquisition trips to Europe to add to the library's collection of books, pamphlets and artwork. Therefore, the majority of materials in this collection come from a Catholic context but we continually aim to steward a more comprehensive collection: Mary has numerous meanings across faith, cultures, traditions and geographical boundaries.

Disrupted Devotion During World War II

Perhaps surprising, given the logistical hurdles involved in long distance travel during times of conflict, Roman Catholic pilgrimage in Europe actually increased during World War II, with many seeking to pray for peace and an end to the war.[6] Others participated in pilgrimage as a way to give thanks for being spared from the conflict, or to appeal to the Virgin Mary and other saints for the safe return of loved ones. An article in the *Catholic Herald* in 1941 described written promises by soldiers mailed to Walsingham and placed under the statue of the Virgin Mary. The soldiers intended to pick up their letters in person if they made it out of the war safely.[7]

Though Catholics are not required to participate in pilgrimage in the same way that Muslims participate in hajj (pilgrimage) as one of the five pillars of Islam, for Christians and more specifically, Roman Catholics, pilgrimage has a long history as a way of expressing devotion. While the nature of pilgrimage changed during World War II from the locales visited, the means of transportation, and even the demographics of those participating, maintaining annual pilgrimages whenever possible was often seen as vital to ensuring the continuation of religious traditions.

For some in Europe, pilgrimage to significant shrines was a reminder of national identity, such as those in Poland who made pilgrimage to the Black Madonna of Czestochowa. However, Kathryn Hurlock argues

6　Kathryn Hurlock, 'Peace, Politics, and Piety: Catholic Pilgrimage in Wartime Europe, 1939–45', *War & Society* 41:1 (2022), 36–52, <https://doi.org/10.1080/07292473.202 2.2021754>.

7　'Soldiers promise Walsingham pilgrimage, if spared', *Catholic Herald*, 10 April 1941, 7.

that 'in England, the maintenance of these traditions was arguably more significant because of the fact that the majority of the population were Protestant, and the establishment and then continuity of these traditions were important in uniting England's Catholic community in a common act.'[8] Some of the struggles of pilgrimage to Walsingham can be found in the publications issued by the two shrines. *Our Lady's Mirror* was created by Fr. Alfred Hope Patten to keep members of the Society of Our Lady of Walsingham in touch with one another. In the winter issue of 1944, it notes, 'owing to the new regulations Walsingham is again in the banned area, and so pilgrimages from outside our own district in Norfolk, Suffolk and Essex are taboo for the present.' The newsletter also mentions inquiries about new pilgrimages but that, 'everything is being held up by the anticipation of the Second Front.'[9]

The people participating in the pilgrimages also changed during the war years. While men of service age were often conscripted as soldiers, it meant that some of the younger generations, including children, were the ones participating. This is one potential reason for shrines broadening their audience for guidebooks in the post-war years. For servicemen and women from the United States, being stationed in Europe gave them an opportunity to visit sites that they might not otherwise have seen. Hurlock explains, 'American airmen and ground crew were so keen to go to Walsingham that they had "forgone their sleep" despite being on duty the night and day before, just so they did not miss out.' A pamphlet in the Marian Library's collection from 1944 details a travel itinerary for a program sponsored by the French government to 'add to the enjoyment of American soldiers on leave in France' with stops at Toulouse, Bordeaux, and Lourdes – home of the Sanctuary of Our Lady of Lourdes, one of the most popular pilgrimage sites in the world.[10]

8   Kathryn Hurlock, 'Peace, Politics, and Piety: Catholic Pilgrimage in Wartime Europe, 1939–45', 36–52.
9   Fr. Alfred Hope Patten, *Our Lady's Mirror* (Winter 1944) <https://www.waaolm. org.uk/olm1944-1.htm>, accessed 31 August 2022.
10   'Tour to Lourdes: This tour is part of a program sponsored by the French government to add to the enjoyment of American soldiers on leave in France'. Paris: Committee for Franco-Allied Goodwill, 1944.

You, Too, Can Go to Europe!

In the years following World War II, there was an increase in travel to several Marian shrines throughout Europe for a few different reasons. Marian devotion in general grew in popularity during the second half of the twentieth century. In 1953 Pope Pius XII declared 1954 a Marian year with the encyclical – a letter from the Pope to bishops about Catholic Church teachings and doctrine – 'Fulgens corona', or 'Radiant Crown'. It was the first Marian year in Church history and, from December 1953 to December 1954, Catholics worldwide were invited to celebrate the Marian year with the purpose 'that the faith of the people may be increased and their devotion to the Virgin Mother of God become daily more ardent'.[11] An article in the *Daily Telegraph* from 16 August 1954 describes a pilgrimage of more than 20,000 to the Roman Catholic Shrine of Our Lady of Walsingham during the Marian year.[12]

There were other reasons for increased Marian devotion in general during this decade. In 1958, the Lourdes shrine in France celebrated its centennial, inviting special pilgrimages and tours to mark the occasion. Many different organizations sponsored group pilgrimages to Lourdes, which often included stops at other Marian shrines throughout Europe. Fr. Philip C. Hoelle, S. M., then director of the Marian Library, led one such tour to Lourdes to mark the occasion. These tours offered opportunities to travel with other like-minded visitors, and the conveniences of packages that included lodgings and transportation meant that pilgrimages during this time period were not necessarily the arduous journeys that required personal sacrifice as in the past.

11   Pius XII. 'Fulgens corona'. The Holy See, 8 September 1953, <https://www.vatican.va/content/pius-xii/en/encyclicals/documents/hf_p-xii_enc_08091953_fulgens-corona.html>.
12   Timothy V. McDonald, *Walsingham 100 Years of Pilgrimage 1897–1997* (Norfolk: Walsingham Centenary Publication, printed by The Lanceni Press Ltd., 1998), 73.

As seen in Figure 9.2, pilgrimages from the United States to Europe were not marketed strictly as religious events, but as once-in-a-lifetime opportunities to see major landmarks and cities in Europe. This particular advertisement was offered by the Catholic Youth Travel Office, based in New York and was sponsored by the National Federation of Catholic College Students (NFCCS) and the National Newman Club Federation (NNCF). The NFCCS was a lay student organization that began in 1937 to raise international, political and social justice issues at Catholic colleges in the United States.[13] Meanwhile, the NNCF was an organization for Catholic students at non-Catholic institutions.[14]

The bottom portion of the flier includes listings for specific shrines to be visited on the different tours, including Our Lady of Walsingham, Our Lady of Fatima, Our Lady of Lourdes with several others. Next to the listing of shrines is a much larger list of 'cities and places' to be visited, including Paris, Madrid and Venice. While many of these European cities do have religious shrines, basilicas and churches, it is interesting that the list of cities to be visited is larger than the list of shrines. Probably due to most Americans' poor knowledge of geography in general, Walsingham is listed as located in London, when in fact it is over 100 miles away.

Pilgrim Pamphlets, But Also for Anyone

For the sake of this chapter, we are looking primarily at mid-twentieth-century printed materials, but the genre of pilgrimage literature is by no means a twentieth-century innovation. Broadly speaking, this genre can communicate the practical details of a physical trip, or it can represent

13   National Federation of Catholic College Students Moderator finding aid, Catholic University of America, <https://libraries.catholic.edu/special-collections/archives/collections/finding-aids/finding-aids.html?file=nfccs>, accessed 18 October 2022.

14   J. D. Long Garcia, 'Newman Centers: a Brief History, U.S. Catholic, <https://uscatholic.org/articles/201201/newman-centers-a-brief-history/> published 16 January 2012.

Figure 9.2: Pilgrimage Poster.

a spiritual journey made by the reader by means of the act of reading; sometimes it does both. An early example of a spiritual pilgrimage text is the thirteenth-century map made by Matthew Paris, showing possible routes from London to Jerusalem and including geographical features, local attractions and native animals.[15] It seems unlikely that the map would have been used to travel physically, but it could facilitate a spiritual trip to the Holy Land. Practical guides for pilgrims to destinations such as Santiago de Compostela were produced at the same period. Another British Library manuscript includes information about water safety and ferry protocols.[16] Like that manuscript, a 1718 publication in the Marian Library, printed in Troyes for French pilgrims, includes songs for pilgrims to sing along their journey.[17]

The genre we refer to as 'pilgrim pamphlets' is typified by a relatively short page count of under fifty pages, contents including some historical information as well as text for trip planning such as lists of hotels, a handful of images which were typically black and white at this mid-century period, often a map including both the pilgrimage destination and also the location of the train station, markets and other practical details. There would also often be some advertisements for other publications or services provided by the publishing body. There are also interesting distinctions among these, particularly in terms of their intended audience. While some are titled as being a guide specifically for pilgrims, others attempt to broaden their audience and appeal, in correlation with other Marian shrines during this time period. What follows are several representative examples of the genre.

Example 1:
Charles G. Mortimer, *Our Lady of Walsingham*. London: Catholic Truth Society, 1948.

---

15   British Library, Royal MS 14 C VII, <https://www.bl.uk/collection-items/matthew-paris-itinerary-map>, accessed 12 September 2022.

16   British Library, Add MS 12213, <https://www.bl.uk/collection-items/santiago-14th-century-pilgrim-travel-guide>, accessed 12 September 2022.

17   *Les chansons des pelerins de S. Jacques: sur l'imprimé à Compostel*. Troyes: 1718. https://udiscover.udayton.edu/permalink/01OHIOLINK_DAYTON/1lnddo/alma991022901149708517.

Length: 28 pages
Price: twopence in 1934, threepence ('3d') in 1948, sixpence ('6d') in 1962

Contents: Primarily historical: Brief introduction, 'The Period of Glory', 'The Period of Destruction', 'The Period of Revival', Conclusion, Appendix of authoritative texts on Walsingham, text on 'The Topography of Walsingham', map of Walsingham.

This pamphlet was first printed in 1934.[18] Following a pause for global conflict, a 'revised edition' was published in 1948. Besides the cover and advertisements, very little of the text changed in subsequent editions from 1951, 1954, 1956 and 1962. Some content shifts around, but for the most part the text remains unchanged. One major addition from 1934 to 1948 is a list of significant events at Walsingham in the intervening years. For example, on 15 August 1934 on the Feast of the Assumption, a Catholic Mass was celebrated in the Slipper Chapel for the first time in 400 years. In 1938, a pilgrimage of 10,000 children came to Walsingham to pray for peace. In May 1945, presumably following Victory in Europe Day, 'there was a large American pilgrimage to the Shrine. There were several other American visits to Walsingham, one with 6 chaplains and 250 men. In July 1946, two exceptionally large pilgrimages took place.'

If the text remained fairly constant for nearly thirty years, could this pamphlet really have facilitated travel? The map, too, remains constant and focuses on the religious site as opposed to the surrounding areas. Our impression is that the frequent reissuing of the pamphlet is itself evidence of its use, and that the text about other significant pilgrimages placed the users of the pamphlet in the company of previous pilgrims. Such a slim volume would be easy to pick up in London – its place of publication – or at Walsingham itself; a traveller might slip it into a pocket either for the journey to Walsingham or as a souvenir for the journey home and away from Walsingham. To this point, the early covers depict the historical 'priory seal' of a seated Virgin and Child, whereas later covers replace that image with a crowd of pilgrims. The other major change from 1934

---

18 1934 edition is available on the Internet Archive, <https://archive.org/details/ourladyofwalsing00omort/page/n1/mode/2up>.

to 1948, and then unchanged in subsequent editions, is the addition of this text to the 'topography' section: 'The grounds are private; Wednesday is visiting day and the fee is sixpence.' This pamphlet is published by the Catholic Truth Society (CTS), a Catholic charity based in London, that was founded in 1868 and remains active today. We will return to it briefly below. The advertisements included in the series of Walsingham pamphlets promote additional CTS publications and where to find them, as well as membership in CTS.

> Example 2:
> Author unknown, *Walsingham: Everybody's Guide Book, with Plans and Illustrations.* Walsingham: The Shrine Office, circa 1960. Printed by R. I. Severs, Cambridge.
> Length: 24 pages
> Price: one shilling

Contents: Architectural and art historical descriptions of the site and its features, interspersed with descriptions of the town. The pamphlet includes photographs, illustrations and a map of Walsingham. This map includes multiple features of the religious site, as well as features of interest to travellers including the railway station, a butcher's shop, High Street shops and the Friday market, a telephone kiosk, the shrine shop, and multiple places one might rest for the night. Advertisements on the back cover direct people to the shrine shop to purchase 'statutes, books, postcards, medals, pilgrim tokens and all types of souvenirs'; The Hospice of Our Lady Star of the Sea for accommodation for pilgrims 'at moderate prices'; and the shrine office for business matters, including the booking of pilgrimages. This publication was intended to facilitate travel, both getting to the site and finding one's way around once there.

Although it has until recently been catalogued with the Marian Library's collections as a publication by Alfred Hope Patten, C.S.A., it varies a great deal from the 1939 first edition digitized by the Walsingham Archives.[19] It shares the cover but little else. It also points out a 'disadvantage

---

19 Fr. Alfred Hope Patten, 'Walsingham: Everybody's Guide Book', 1939. Published by the Shrine Crafts Shop, Walsingham, Norfolk and printed at Seeley's Printing Works, Wells, Norfolk. Digital photographs provided by the Walsingham Archives and hosted here: <https://www.waapublications.org.uk/wegb.htm>.

to [Patten's] theory' about some archaeological points, suggesting another author who was nevertheless very familiar with Patten's work. Patten, an Anglo-Catholic priest, was instrumental in work to restore Walsingham and promote pilgrimages to the site, sometimes facing significant opposition in the process. The pamphlet's subtitle, 'everybody's guide book', gestures to ecumenism. As a publication of the Church of England shrine office, it suggests that the pamphlet and by extension the shrine, are for everybody.

In the upper right corner of the inside title page is the inscription, 'R. Maloy'. Fr. Robert Maloy, S.M., became the acting director of the Marian Library in 1967 after returning from his doctoral studies in Europe. Many of the Marian Library's materials made their way into the collection based on what members of the Society of Mary had access to through their travels or through their personal and professional networks. Given that this pamphlet was published sometime after 1959, it's very possible that it was Fr. Maloy's personal copy that he had used on a trip to Walsingham and then later donated to the collection.

> Example 3:
> Horace Arthur Bond and Claude Fisher, *The Story of Our Lady of Walsingham with Pictures Guide and Plans*. Walsingham: Greenhoe Press, 1950.
> Length: 20 pages
> Price: one shilling

Contents: Sections on 'The Story', 'Guide for Visitors' and 'Archaeological Notes'. Two maps are also included, one of the local Walsingham area and another of the Walsingham Augustinian Priory. Twelve black and white photographs are interspersed throughout the text, featuring the grounds, a procession from a youth pilgrimage and other ceremonies and points of interest. The back cover includes advertisements for the Walsingham Catholic Pilgrim Bureau, Guildshop Restaurant ('Breakfasts, Lunches, Suppers, Teas & Light Refreshments, parties of up to 250 catered for') and St Mary's College & Chorister School for boys, admitted from the age of 7 years.

Although the title of this pamphlet suggests that it is mostly historical information about Our Lady of Walsingham, this guide includes a sizable section aimed at visitors. The descriptions serve as a sort of walking tour with numbered points of interest featured on the accompanying map. Visitors are given very specific directions to navigate from one point to the next, such as, 'turn right as you leave the gatehouse. A few paces to your left and opposite the old village conduit house [8], with its cresset top, is the building formerly the Swan hostelry [7].' This text brings to mind almost a pilgrimage within the primary pilgrimage, as the pamphlet guides visitors through their experience of the space.

Greenhoe Press, unlike the Catholic Truth Society, was actually situated in Walsingham. The shrine appears to have been its main publication topic. A WorldCat search reveals eight titles, some of which are different editions of this title. The same publisher appears also to have published postcards of Walsingham, which could also be an important piece of the intersections between tourism and pilgrimage–both the sending of postcards to friends at home and the purchase of them as mementoes.

> Example 4:
> *The pilgrims' guide (with history, prayers, hymns and pictures) to the shrine of Our Lady of Consolation, West Grinstead, nr. Horsham, Sussex.* Faversham: Kent Carmelite Press, circa 1951.
> Length: 24 pages
> Price: one shilling

Contents: Historical information about Our Lady of Consolation and the West Grinstead pilgrimage. Also includes several photographs and hymns, including one specific to Our Lady of Consolation at West Grinstead. The catalogued pamphlet also included three blank postcards – two from the shrine and one featuring an image of Our Lady of Consolation from Turin, Italy.

This pamphlet is particularly interesting given its evidence of ownership by someone, though there is no clear provenance. The publication date is not clear, with the catalogue record suggesting '1951?' An earlier version from 1948 is also available in the Marian Library's collection, with some slight variations in the title and contents. The title page of this version includes a pasted in note updating information for the Annual Pilgrimage and the Children of Mary's Pilgrimage. This pasted note even includes the specific details of, 'Procession, 2.45. High Mass, 5.15 p.m'. Presumably then, this pamphlet is intended to be more ephemeral, useful for a visit today, but perhaps outdated tomorrow.

We include this non-Walsingham as an example that, while Walsingham certainly took the lead for pilgrimage fervour in this period, it was not the only destination for Marian devotion. Notably, although the shrine at West Grinstead could only accommodate 100 people, marking it as much smaller in scale that Walsingham, its pamphlet is of a similar print quality and is offered at approximately the same price as Walsingham publications from the same period.

Publication, Distribution, Conclusions

Until the early nineteenth century, printed matter such as books, periodicals and newspapers, were produced by a series of labour-intensive manual processes. Typesetting, engraving, paper-making, printing and bookbinding, are among the many processes that led to the creation of printed matter for distribution. The materials produced in this era were, understandably, more expensive and less readily available to the wider public, although there were nevertheless cheaper products available even in the letterpress era. Things changed with the introduction of new machines and methods that made publication cheaper and faster, for example steam-operated printing machines. The industrial revolution in publishing impacted every field of knowledge in its own way, from the explosion of popular literature for a general audience (novels sold in instalments to be read on passenger trains, for example) to the broad distribution of botanical literature, leading to various trends in non-elite adoption of gardening and horticulture.[20] Naturally, it would also impact the production and distribution of religious literature.

20 Kevin J. Hayes, 'Railway Reading', *Proceedings of the American Antiquarian Society* 106: 2 (1997), 301–26. <https://www.americanantiquarian.org/proceedings/44525120.pdf>; See, for example, the popular works by John C. and Jane Loudon and the popularity of *Curtis's Botanical Magazine*, which is still in publication.

The barriers to entry, both for publishers and for readers, within and beyond Britain, were lowered at just the right time to coincide with other events discussed above, namely the rise of Anglo-Catholicism and the Oxford Movement. Also relevant was the papal encyclical 'Ineffabilis Deus', distributed in 1854 and defining the Roman Catholic dogma of the Immaculate Conception, the belief that the Virgin Mary was born free of original sin. This encyclical concluded stirringly:

> With a still more ardent zeal for piety, religion and love, let them continue to venerate, invoke and pray to the most Blessed Virgin Mary, Mother of God, conceived without original sin. Let them fly with utter confidence to this most sweet Mother of mercy and grace in all dangers, difficulties, needs, doubts and fears. Under her guidance, under her patronage, under her kindness and protection, nothing is to be feared; nothing is hopeless. Because, while bearing toward us a truly motherly affection and having in her care the work of our salvation, she is solicitous about the whole human race. And since she has been appointed by God to be the Queen of heaven and earth and is exalted above all the choirs of angels and saints, and even stands at the right hand of her only-begotten Son, Jesus Christ our Lord, she presents our petitions in a most efficacious manner. What she asks, she obtains. Her pleas can never be unheard.[21]

This confirmation of what many Catholics already believed, followed four years later by the founding apparitions of the present-day Sanctuary of our Lady of Lourdes, France, gave ample reason for Roman Catholics and other devotees to pay personal visits to Marian shrines. The Catholic Truth Society (CTS), founded in London in 1868 'to help the Catholic community living in the shadows, persecuted and neglected', produced a huge variety of literature, many of which were pamphlets such as those listed above, including brief guides to pilgrimage sites and short lives of Catholic saints.[22] CTS remains active today, with an international reach; today they advertise bulk orders of 'leaflets' such as *Why Should I go to Mass on Sunday?* via their website. From its earliest days, the CTS distributed

21 Pope Pius IX. 'Ineffabilis Deus'. The Holy See, December 8, 1854, <https://www.papalencyclicals.net/pius09/p9ineff.htm>.

22 'Our History', Catholic Truth Society, <https://www.ctsbooks.org/about-us/our-history>, accessed 7 November 2022.

pamphlets, some sold or given away near religious sites and others mailed out via subscription. The small size of the pamphlet as a genre ideally suited it for a pocket or a pocketbook, for reading on the train or over lunch, for a physical pilgrimage or a spiritual one.

The Catholic Truth Society was only one distributor of pilgrim pamphlets, however. Taking as examples the pamphlets listed about, we see a much smaller scale in the model of Greenhoe Press, which was based in and published exclusively about Walsingham. Splitting the difference, the Kent Carmelite Press – located about eighty miles (130 kilometers) from Our Lady of Consolation, West Grinstead – appears to have been run by the British Province of Carmelites, an order with a particular devotion to the Virgin Mary as Our Lady of Mount Carmel. A WorldCat search demonstrates that most of their publications were on their own religious order, but there must have been regional interest in the revived devotion to the destroyed medieval shrine which was rebuilt in the late nineteenth century.

The one Anglo-Catholic example provided above lists the publisher as the shrine office'; however, it was in fact printed seventy miles (113 kilometers) away, in Cambridge; the printer, R. I. Severs, published non-religious materials such as postcards of the university, as well. The genre we discuss in this chapter is alive and well. The same person who visits a site to pray, also needs a place to sleep, also purchases a trinket to bring home, and so on. The person who begins as a tourist may reconsider their faith life in the context of their trip. We have heard anecdotes of devout individuals who have been 'turned off' by the commercialism of religious sites marketing trinkets on a large scale; so, pilgrimage can even be a site of disillusionment! The pilgrim becomes a jaded tourist. Pilgrims might 'visit' a shrine remotely, via a paper pamphlet or through the internet, bookmarking the web page of a favourite shrine. The means of publication and dissemination have simply increased, however, rather than replaced the methods documented in this chapter; pamphlets remain an abundant product of Christian religious life. Pilgrimage and tourism are interconnected and have generated, and will continue to generate, a rich body of material culture that records and demonstrates how people and holy sites interact with one another.

10 'Nothing more than my own personal notebook': The Composition, Design and Consumption of Alfred Wainwright's *A Pictorial Guide to the Lakeland Fells*[1]

This chapter places Alfred Wainwright's individual approach to the composition and design of his *A Pictorial Guide to the Lakeland Fells* in the context of selected preceding guidebooks to the English Lake District. It begins by introducing Wainwright and the guidebooks which he described as 'nothing more than my own personal notebook'.[2] It considers idiosyncratic aspects of his method of making the *Pictorial Guides* and their distinctive nature before drawing attention to ways in which these guidebooks reflect certain themes and tensions expressed in their precursors. The chapter concludes by suggesting that the combination of Wainwright's personal approach with elements drawn from English Lake District print history contributed to the *Pictorial Guides* being received both as superlative

1   I would like to thank the audience at the Centre for Printing History & Culture (CPHC) Printing for Tourists Conference held in July 2020 for their illuminating responses to an earlier version of the work which informs this article. The CPHC awarded me the 2022 Peter Isaac essay prize which has allowed me further to develop my ideas. I am grateful to the Estate of A. Wainwright for granting me permission to reproduce text and illustrations from the *Pictorial Guides*; Derry Brabbs for his permission to reproduce the photograph of Alfred Wainwright in Figure 10.1; Derek Cockell, David Johnson, Barry McKay, Timothy Sykes and Sheila Wiggins for sharing their knowledge, and Simon Bainbridge and Christopher Donaldson, my PhD supervisors at Lancaster University, for their enthusiasm for my research.

2   Hunter Davies, ed., *The Wainwright Letters* (London: Frances Lincoln, 2011), 95. Unsent letter to Bill Mitchell, editor of *Cumbria* magazine.

examples of illustrated walkers' guidebooks and as transcending the form of consumption typically represented by 'printing for tourists'.[3]

Many people view Wainwright as the godfather of modern fell-walking. The Oxford English Dictionary defines the latter as a practice of walking over rough ground connected with a particular landscape, chiefly that of the mountainous country or fells of the north-west of England centred on the Lake District National Park (LDNP) and parts of Scotland. Wainwright's pre-eminent status has largely arisen from the reception of his seven-volume *Pictorial Guides*, first published between 1955 and 1966. His description of the *Pictorial Guides* as 'nothing more than my own personal notebook' reflected their method of composition, design and artisanal appearance in print. Atypically for illustrated walkers' guidebooks designed as practical volumes to be used in the field, they also have been received as works of art and literature to be enjoyed by the armchair reader.

Handling one of the seven volumes of the *Pictorial Guides* today it is easy to take for granted the success of Wainwright's idiosyncratic approach to composition and design. In an interview in 1982 he stated that his method 'suited my particular circumstances and my own idea of the sort of book I would have liked for myself'.[4] Wainwright created the *Pictorial Guides* through thirteen years of personal research and exploration of the 214 fells he chose to include.[5] The *Pictorial Guides* were printed but appeared to be entirely written and drawn by hand. Wainwright used photographs taken in the field to compose his drawings, but photography was not used in his published guidebook. His approach contrasted with twentieth-century guidebooks such as those by the Abraham Brothers and W. A. Poucher, which depicted the mountain landscape of the district through the increasingly popular medium of photography. Wainwright

3 Wainwright's contemporary and friend, the author and journalist A. Harry Griffin, called *Book One, The Eastern Fells*, 'the most remarkable book of its kind about the Lake District ever printed' in a review in the *Lancashire Evening Post*, 27 May 1955.
4 Ken Garland, 'Lead, Kindly Light, A User's View of the Design and Production of Illustrated Walkers' Guides', *Information Design Journal* 7/1 (1993), 55.
5 Hunter Davies, *Wainwright The Biography* (London: Michael Joseph, 1995), 134.

did not conduct any research with his potential readership regarding the format he adopted, took no account of printing methods at the time, and had no publisher in place as Book One neared completion.[6]

Wainwright described the design of his personal notebook as common sense.[7] The *Pictorial Guides* are pocket-sized to facilitate ease of use on the fells. Each fell has a separate chapter with a consistent structure: an introduction to and illustration of the fell from valley level; a simple plan view pinpointing its location; detailed maps; illustrations and text describing its notable features; ascent diagrams, such as the Ascent of Blencathra from Threlkeld shown in Figure 10.1; a description and drawings of the summit and summit views and panoramas, ridge routes and notes on routes of descent.[8] Wainwright integrated text and over 3,600 of these illustrations in a consistent style throughout the series.[9]

The rationale for Wainwright's handcrafted design choices was personal – emotional, pragmatic and rooted in the transformative effect of his first visit to the English Lake District in 1930. He described the *Pictorial Guides* as a tribute or 'love letter' to Lakeland when Book One was published twenty-five years later.[10] Factors which may have been constraints for other authors played to Wainwright's abilities and interests. His methodical and handcrafted composition and design process united

6   Alfred Wainwright, *Memoirs of a Fellwanderer* (London: Michael Joseph, 1993; repr. Frances Lincoln, 2003), 80–2.

7   Clive Hutchby, *The Wainwright Companion* (London: Frances Lincoln, 2012), 24–5. Hutchby references a talk by Ken Garland to the British Cartographic Society's Annual Technical Symposium at Reading University in 1996 entitled 'Passionate Physiographer: The Design and Execution of A. Wainwright's Pictorial Guide to the Lakeland Fells'.

8   Alfred Wainwright, *A Pictorial Guide to the Lakeland Fells, Book Five, The Northern Fells* (Kentmere: Henry Marshall, 1962), Blencathra 17–8. It may be helpful to explain my referencing of the *Pictorial Guides* here. These guidebooks are divided into seven volumes by geographical area. The fells are arranged in alphabetical order in each volume, with page numbers which start from one for each fell and also bear its name. The numbering of the next fell in the book restarts from one.

9   Hutchby, *The Wainwright Companion*, 18.

10 Wainwright, *A Pictorial Guide, Book One, The Eastern Fells* (Kentmere: Henry Marshall, 1955), Introduction.

A PICTORIAL GUIDE
TO THE

LAKELAND FELLS

being an illustrated account
of a study and exploration
of the mountains in the
English Lake District
by

a wainwright

BOOK ONE

THE EASTERN FELLS

Figure 10.1: Alfred Wainwright, the *Pictorial Guides* and 'Ascent of Blencathra from Threlkeld', *A Pictorial Guide, Book Five, The Northern Fells*, Blencathra 17 © Derry Brabbs; The Estate of A. Wainwright.

his professional sensibility as an accountant trained in handwritten ledgers with his skill in illustration, and also made use of the materials he had available. Graphic designer and author Ken Garland has, however, pointed out some disadvantages of Wainwright's approach. For example, Wainwright's ignorance of printing techniques made his method of correcting mistakes by redrafting entire pages more laborious than necessary.[11]

It would be a mistake to assume that the initial and continued success of the *Pictorial Guides* was certain. The *Pictorial Guides* demand that the reader-walker commits more effort in engaging with the text, illustrations and the author than is typical of modern guidebooks. Wainwright's maps are his own design and adopt conventions different from official versions. His opinions are personal and do not aim for inclusivity. Tourists, motorists and particular sorts of women are given short shrift. Selecting a walk and planning a day on the fells which includes more than the ascent and descent of one summit necessitates the reader-walker connecting one peak to the next by consulting the ridge routes section of the relevant chapters. Apart from a very few exceptions, circular or linear walks are not suggested.[12] Wainwright's omission of such routes supports the design of the *Pictorial Guides* as a personal notebook, albeit one which was intended to encourage his readers' own explorations. It also suggests that he assumed a certain level of competence in users of his books. Wainwright did not intend his guidebooks to be updated in his lifetime and recommended as early in the series as Book Three, *The Central Fells*, that his readers use them as the basis for their own, updated, notebooks.[13]

11 Garland, 'Lead, Kindly Light', 47–66, 54–5. Wainwright drew each page by eye using dip pen and ink, to the size of its intended reproduction in print, on one sheet of paper with no cutting and pasting. He discarded the first ninety pages of Book One because he decided that the text would look neater if it were justified on the right-hand side as well as the left.

12 See Wainwright, *A Pictorial Guide, Book One*, Fairfield 3; *Book Six, The North Western Fells* (Kendal: Westmorland Gazette, 1964), Causey Pike 10.

13 Wainwright, *A Pictorial Guide, Book Three, The Central Fells* (Kentmere: Henry Marshall, 1958), Some Personal Notes in Conclusion.

Differences between the *Pictorial Guides* and modern guidebooks suggest that the highly personal nature of the former may have prevented Wainwright's design from being universally adopted. Modern guidebooks typically isolate any information extraneous to the walking route directions in a separate text box and illustrate the walk itself with colour photographs.[14] In the *Pictorial Guides*, text and illustrations are in black and white and interspersed with textual guidance and the author's opinions. The reader-walker may find that the *Pictorial Guides'* seven-volume design mitigates against compactness and is at odds with the single volumes of most of Wainwright's precursors and successors. Tourists based, for example, at the head of the Great Langdale valley would require three volumes of the *Pictorial Guides* (*The Southern Fells*, *The Central Fells* and *The Western Fells*) to enjoy all the walks accessible from the door of their accommodation. Today's guidebooks frequently make route choice easier for their readership by presenting a selection of best walks based on criteria such as the time available, distance covered and arduousness of the terrain encountered.[15] Instead, Wainwright recommended best routes of ascent and descent of individual fells rather than the best walks. The reader-walker had to wait eleven years from the publication of Book One in 1955 for Wainwright to reveal his various 'best' lists in the conclusion to Book Seven which was published in 1966. There are, however, examples of the influence of Wainwright's design choices upon modern guidebooks to the English Lake District and hand-drawn maps to other areas of the country. Mark Richards's series of *Fellranger* guidebooks are arranged as a curated collection of fells divided geographically and listed in alphabetical order in each volume, echoing the structure of the *Pictorial Guides*, and include what Richards describes as linescape drawings and fellscape diagrams.[16] Jack Keighley made hand-drawn and illustrated guides to walks

14 I am grateful to Joe Williams, Business Development Manager, Cicerone (UK walking, cycling and mountain guide publishers) for this insight given in a telephone interview on 18 February 2021.
15 See for example Mark Richards, *Great Mountain Days in the Lake District, 50 Classic Routes Exploring the Lakeland Fells* (Milnthorpe: Cicerone, 2008).
16 Mark Richards, *Lakeland Fellranger, The Far Eastern Fells* (Kendal: Cicerone, 2013).

in Lancashire, Yorkshire and the Pennines.[17] Many villages in the county of Suffolk have hand-drawn footpath maps which amateur map maker Wilfrid George started to produce in the 1960s. Recently, Colin Hindle has composed two hand-drawn and handwritten Lakeland guidebooks in a planned series of four. These depict the best Lakeland views and are intended to be supplemental to the *Pictorial Guides*.[18]

Why then have the *Pictorial Guides* remained in continuous use for over seventy years? Studies in the field of graphic design have suggested that their popularity is attributable to aspects of their nature as a personal notebook. Angharad Lewis has summarized the *Pictorial Guides* as examples of the 'most human side of superlative information design'.[19] Garland has suggested that the handwritten and hand-drawn nature of Wainwright's work possesses an advantage over print by engendering trust in the author as guide and companion. This advantage only applies if the handwritten guide is executed well. Garland has drawn attention to the coherence of text and illustration, which enables ready cross referencing, the accuracy and appropriateness of the information included, and its intuitive presentation, all being the elements which ensured Wainwright's successful communication of information to his readership.[20]

Three features, relating to illustration, of the way in which Wainwright set about what he described as 'building a mountain on a blank sheet of paper' in the *Pictorial Guides* contributed to generating such trust and confidence in the reader-walker that it seemed as if the author were accompanying them on their explorations.[21] Wainwright's innovative ascent diagrams,

17 See for example Jack Keighley, *Walks in the Yorkshire Dales: An Illustrated Guide to Thirty Walks of Outstanding Beauty and Interest* (Milnthorpe: Cicerone, 1989).

18 Colin Hindle, *The Best Lakeland Views: A hand drawn, hand written guide to the Lake District, Book One Ullswater, Brotherswater, Haweswater* (Penrith: H&H Reed, 2021); *Book Two Windermere, Grasmere, Langdale, Coniston* (Penrith: H&H Reed, 2024).

19 Angharad Lewis, 'Drawn to the Land', *Eye: The International Review of Graphic Design*, 20.78 (2010), 48–9.

20 Garland, 'Lead, Kindly Light', 54–8.

21 Alfred Wainwright, *Fellwanderer, The Story Behind the Guidebooks* (Kendal: Westmorland Gazette, 1966), fol. 4v.

maps and self-portraits combine with other topographical illustrations to lead the reader-walker through the landscape, varying viewpoints and scale dynamically to zoom in and out on features of interest and importance to the walker on the ground.[22] Clive Hutchby has pointed out that the imaginary viewpoints hovering above the ground in Wainwright's ascent diagrams (shown by the example in Figure 10.1, 'Ascent of Blencathra from Threlkeld' via Hall's Fell) gave a walker's eye view of the terrain extending beyond the summit in a way that a photograph, such as those in Poucher's pocket guidebook *The Lakeland Peaks* (1960), could not.[23] Wainwright's practical experience as a fell-walker and his passion for reading and drawing maps led him to pinpoint the deficiency of mapping in respect of footpaths for walkers in the maps of the Ordnance Survey (OS) and Bartholomew.[24] Accurate maps were of such significance to him that his hand-drawn versions appeared throughout the text of the *Pictorial Guides*.[25]

Wainwright varied the scale of his maps according to the complexity of the terrain he wished to represent.[26] The three types of footpath shown in the map key in Figure 10.2 were intended better to inform the walker on the ground. By contrast, some of the legally designated rights of way shown on OS maps had fallen into disuse, whilst other well-trodden paths were omitted.

Finally, Wainwright's self-portraits reinforced his role as guide and had the effect of moving his readership away from the passive consumption of a printed object to being in the landscape more actively with the author. His self-portraits visually amplified his own status as guide and companion. He typically drew himself in close-up and other figures in miniature. He

22 Liz Woodham, '"Building a Mountain on a Blank Sheet of Paper": Alfred Wainwright's A Pictorial Guide to the Lakeland Fells', *Publishing History* 87 (2023), 7–38 examines these aspects of the Pictorial Guides in more detail.

23 Hutchby, *The Wainwright Companion*, 26–7. W. A. Poucher, *The Lakeland Peaks* (London: Constable, 1960).

24 Wainwright, *A Pictorial Guide, Book One*, Introduction, Notes on the Illustrations.

25 Wainwright, *A Pictorial Guide, Book Five*, Skiddaw 9.

26 Derek Cockell, 'Wainwright on Cartography', *Footsteps*, Wainwright Society Magazine, 76 (2022) 20–3.

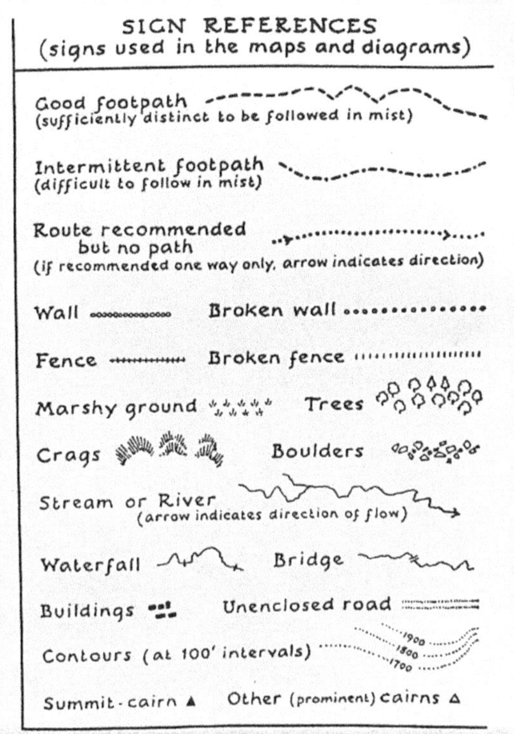

Figure 10.2: A Wainwright map, *A Pictorial Guide, Book Five, The Northern Fells,* Skiddaw 9. © The Estate of A. Wainwright.

placed his self-portraits in locations which were either significant to him personally or important for the reader-walker to get right, such as The West Wall Traverse on Scafell in Book Four shown in Figure 10.3.[27]

Wainwright's description of himself as the Oracle is simultaneously ironic and self-consciously emblematic. His correspondence includes a letter thanking a reader for a photograph which he praised for its effective and symbolic arrangement of himself against a cairn and the sky.[28]

27 Wainwright, *A Pictorial Guide, Book Four, The Southern Fells* (Kentmere: Henry Marshall, 1960), Scafell 9.
28 Davies, *Letters*, 115. I am indebted to Derek Cockell for bringing this to my attention.

Figure 10.3: The Oracle. © The Estate of A. Wainwright.

The striking ways in which the appearance of the *Pictorial Guides* diverged from previous English Lake District guidebooks points to Wainwright's aim for his work to take a place among its precursors in the print history of the district. His early correspondence with former work colleagues, after he had moved to Kendal, adjacent the Lake District, in 1941, referred to his desire to write a guidebook and for his name to be remembered alongside those of the Lake Poets – William Wordsworth, Robert Southey and Samuel Taylor Coleridge.[29] Wainwright's aspirations for the *Pictorial Guides* seemingly prioritized meeting his aims for them to rank in quality alongside the pantheon of Lakeland writers and echoed how earlier

29 Davies, *Letters*, 51–70.

guidebook authors such as Harriet Martineau had consciously located themselves on the literary map of Lakeland alongside Wordsworth.[30] In practical terms Wainwright recognized two guidebooks as precursors to the *Pictorial Guides*. He stated that after his first visit he came to know by heart both a map of the district and a second-hand copy of M. J. B. Baddeley's *Thorough Guide to the English Lake District*.[31] The *Thorough Guide* was one of the most widely read and influential guidebooks to the area based on longevity alone. First published in 1880 during the heyday of late Victorian pleasure travel, it remained in print for nearly a century (the twenty-sixth and final edition was published in 1978). After he had completed the *Pictorial Guides*, Wainwright referred to Henry Irwin Jenkinson's *Practical Guide to the English Lake District* (1872) as the best guidebook to the district yet written.[32]

A consideration of the *Pictorial Guides'* full title – *A Pictorial Guide to the Lakeland Fells being an illustrated account of a study and exploration of the mountains in the English Lake District* – illuminates significant ways in which Wainwright's personal notebook was in dialogue with, as well as being distinctive from, its precursors. The Lakeland Fells and English Lake District of the *Pictorial Guides'* title reflects a print tradition which increasingly brought the uplands of the district into focus and contributed to the formation of the identity of the area. The guidebook widely recognized as being the first to the district, Thomas West's *A Guide to the Lakes: In Cumberland, Westmorland, and Lancashire* (1778) largely described a landscape of valleys and lower elevations as well as lakes.[33] The *Pictorial*

30 Alexis Easley, *Literary Celebrity, Gender and Victorian Authorship 1850–1914* (Lanham: University of Delaware Press, 2011), 50–66.
31 Wainwright, *Fellwanderer*, fol. 4 v; M. J. B Baddeley, *Thorough Guide Series, The English Lake District* (London: Dulau, 1880).
32 Alfred Wainwright, *Wainwright's Favourite Lakeland Mountains* (London: Michael Joseph, 1991), 1; Henry Irwin Jenkinson, *Jenkinson's Practical Guide to the English Lake District with Maps* (London: Edward Stanford, 1872).
33 Thomas West, *A Guide to the Lakes: Dedicated to the Lovers of Landscape Studies, And to All Who Have Visited, Or Intend to Visit the Lakes In Cumberland, Westmorland and Lancashire* (London: Richardson and Urquhart; Kendal: W. Pennington, 1778).

Guides represent more active engagement with the landscape of the district from a height. Wainwright's imaginary and mobile perspectives are the direct opposite of West's stations or fixed viewing points from where the landscape could be admired as if in a picture. The naming of the area itself has changed from its component counties in West's time to the developing concept of Lakeland and the Lake District or English Lake District as it increasingly became known in nineteenth- and twentieth-century guidebook literature. This change is seen both in the title of Wainwright's *Pictorial Guides* and their structure. The latter focused on the detailed exploration of the individual fells in one of seven upland areas defined by the author within the boundaries of the new LDNP which had been carved out of the central high ground of the English Lake District in 1951, four years before the publication of Book One.

The fell character studies which are the core of Wainwright's *Pictorial Guides* represent a culmination of guidebook content which, from the late eighteenth century, increasingly informed and reflected pedestrian exploration of the uplands of the district for pleasure. Ann Radcliffe's account of climbing Skiddaw in her well-received *Observations during a Tour to the Lakes of Lancashire, Westmoreland, and Cumberland* (1795) was included in the Addenda to all editions from the sixth (1796) onwards of West's *Guide to the Lakes*.[34] Ascents of popular mountains such as Skiddaw and Helvellyn, whether on foot or horseback, frequently formed part of domestic tour itineraries. Late eighteenth- and early nineteenth-century guidebook writers commonly warned their readers not to take such upland explorations lightly, advising them to hire a local guide and quoting poems by Wordsworth and Sir Walter Scott which commemorated the death of Charles Gough in a fall on Helvellyn in 1809 and his faithful dog's vigil beside his master's body.[35] Other authors recorded their exploits on the fells in thrilling terms. Edward Baines represented his ascent of Striding Edge

---

34 Penny Bradshaw ed., *Observations During a Tour to the Lakes of Lancashire, Westmoreland and Cumberland*, Ann Radcliffe, (Carlisle: Bookcase, 2014), 39.

35 Recounted in Rev. H. D. Rawnsley, *The Story of Gough and His Dog on Helvellyn* (Carlisle: G&T Coward, 1892). Rawnsley organized the placing of a monument to Gough on Helvellyn in 1890.

on Helvellyn, unusually without a guide, as daring in his *Companion to the Lakes of Cumberland, Westmoreland, and Lancashire* of 1829.[36]

Developments in cartography and its representation in print helped to alter the conventional advice in early guidebooks not to venture alone into the fells of the district without hiring a guide, effectively a form of servant. The inclusion of maps and detailed walking route directions in later nineteenth-century guidebooks facilitated the opening of pedestrian exploration of the uplands of the English Lake District to new types of visitors. Harriet Martineau unusually encouraged the reader of her *A Complete Guide to the English Lakes* (1855) to experience 'A Day on the Mountains' before leaving Westmorland for Cumberland, alone if possible, to experience elevated views and the solitude of the hills.[37] She chose Fairfield as a subject for exploration on the grounds of its safety in good weather. She recommended elsewhere in her guidebook that a guide was hired for upland expeditions to avoid accidents. Thirteen years after Martineau's guidebook was first published, the completion of the OS mapping of Westmorland and Cumberland in the six-inch-to-one-mile scale formed a watershed moment, as accurate maps of the region's complex landscape changed what was expected of a guidebook. The maps and route directions for walkers in Wainwright's guidebooks had notable and popular precedents in Herman Ludolph Prior's *Ascents and Passes in the Lake District of England* (1865), Jenkinson's *Practical Guide* and Baddeley's *Thorough Guide*.[38] These guidebooks typically included fold-out and sectional maps of the district and expanded the list of ascents beyond those most well-known. The cartography in Baddeley's *Thorough Guide* was particularly important. The first use of printed layer colouring to denote contours rising in 500 feet increments up to 2,500 feet

36 Edward Baines, *A Companion to the Lakes of Cumberland, Westmoreland, and Lancashire in a Descriptive Account of a Family Tour, and an Excursion on Horseback: Comprising a Visit to Lancaster Assizes. With a New, Copious, and Correct Itinerary* (London: Hurst, Chance & Co., 1829), 215-16.

37 Harriet Martineau, *A Complete Guide to the English Lakes* (Windermere: Garnett, 1855; London, Whittaker, 1855), 57–65.

38 Herman Ludolph Prior, *Ascents and Passes in the Lake District of England* (London: Simpkin, Marshall; Windermere: Garnett, 1865).

helped his readers better to understand the height and potential difficulty of the landscape.[39] Such maps enabled the sort of adventurous walking represented by ascents of the exposed, rocky and airy ridges of Striding Edge on Helvellyn and Sharp Edge on Blencathra (as well as less challenging expeditions) to come, by the later nineteenth century, within the reach of visitors without the means to hire a guide.[40] Further developments in cartography assisted the sharpening of the focus on the uplands of the district represented in the *Pictorial Guides*. Wainwright acknowledged the importance to his guidebook project of the re-publication of the two-and-a-half-inch OS maps from 1937–61 which enabled him to make his own accurate illustrations and cartography.[41]

Wainwright's emphasis on *Pictorial* in his title echoed the focus present from early guidebooks on appreciating the picturesque scenery of the district, described in Peter Bicknell's authoritative bibliography.[42] Wainwright's depictions of summit views, panoramas and landscape features can be traced in early mountaineering literature.[43] Subsequent developments in nineteenth-century print enhanced the portrayal of the uplands in guidebooks. The *Thorough Guide* and the *Practical Guide* contained detailed lithographs of panoramas of a range of summits, (included in Jonathan Otley's *A Concise Description of the English Lakes, and Adjacent Mountains* since the fourth edition of 1830, but in less detail), while the importance Jenkinson placed on the naming of the fells is echoed in the *Pictorial Guides*.[44] The increasing importance of illustrations points to

39   Leslie Gardiner, *Bartholomew: 150 Years* (Edinburgh: Bartholomew, 1976), 106.
40   See for example Herman Ludolph Prior, *Prior's Guide to the Lake District of England*, 3rd edn (Windermere: Garnett [1881–5]), 84–5; 130.
41   Wainwright, *Memoirs of a Fellwanderer*, 69–70.
42   Peter Bicknell, *The Picturesque Scenery of the Lake District 1752–1855: A Bibliographical Study* (Winchester: St Paul's Bibliographies, 1990).
43   Simon Bainbridge, 'Romantic Writers and Mountaineering', *Romanticism* 18/1 (2012): 1–15; 6.
44   Barry McKay, 'Thoroughly Baddeley', lecture to the Wordsworth Trust (2015). The fifth edition of the *Thorough Guide* in 1889 includes a panorama of the fells in three illustrations.

a potential conflict between beauty and utility in guidebook design, suggested by the positioning of maps and other illustrations as imported accessories in later nineteenth-century guidebooks and Poucher's superimposition of handwritten text and directional arrows on his photographs in early editions of *The Lakeland Peaks*.[45] The greater visual coherence of Wainwright's handwritten and hand-drawn design and his integration of illustration and text largely resolved these issues. His drawings can be appreciated in their own right and help to bring the reader-walker into closer acquaintance with the landscape.

Wainwright's description of the *Pictorial Guides* as an 'account' and 'study' points both to his methodical composition process and the increasingly scientific emphasis James Buzard has identified in earlier guidebooks in *The Beaten Track*, his analysis of nineteenth-century European tourism and culture.[46] Buzard has described how the guidebook industry as a whole was influenced by the production of guidebooks to the Alps and other parts of the world as improvements in transport heralded the dawn of the age of mass tourism. In the English Lake District, the arrival of the railway in 1847 in what was Birthwaite and was renamed Windermere brought increasing numbers of more diverse visitors to the region. Martineau's *Complete Guide* was written with such visitors in mind and aimed to provide those unfamiliar with the district with practical, objective and comprehensive information.[47] Wainwright's record of his explorations in the *Pictorial Guides* formed a study marked by a methodical approach to providing similar information on the uplands of the English Lake District. His fell character studies touched on aspects of preceding guidebooks as well as the personal; features that caught his eye and spoke

45 See for example Poucher, *The Lakeland Peaks*, 222. Route 71 depicts the Ascent of Blencathra via Hall's Fell.

46 James Buzard, *The Beaten Track: European Tourism, Literature and the Ways to 'Culture', 1800–1918* (Oxford: Oxford University Press, 1993), 65. Buzard describes the standardization, constant updating, diligence and thoroughness which Murray and Baedeker brought to the guidebook, 'making it not the record of someone's tours but a description of what current tourists could anticipate'.

47 Martineau, *Complete Guide*, 3.

to his interests. The resulting diverse content supports Paul Readman's description of the way in which the literature of the district helped to develop an understanding of its landscape which was characterized by past and present human interactions.[48] Wainwright paid attention to cairns, memorials, relics of abandoned industries and the well-crafted sheepfolds of Skiddaw Forest alongside his explanations of topographical curiosities and watersheds.[49] The *Pictorial Guides* were informed by the Scandinavian and Roman history of the area, reflected notably in Jenkinson's and Baddeley's guidebooks and the writings of W. G. Collingwood in the late nineteenth century. Wainwright also included literary references, primarily to locations in Wordsworth's poetry.[50] These references enabled visitors to experience places associated with the life and work of the poet as had earlier guidebooks, facilitated by the improvements in rail and motor transport to the district described by Saeko Yoshikawa and Jean Turnbull.[51]

'Exploration of the mountains' in the *Pictorial Guides* brought the history of pedestrian exploration of the fells in preceding guidebooks into a contemporary context. Wainwright noted reading and re-reading mountain literature including works by Frank S. Smythe and other classics,

48 Paul Readman, *Storied Ground, Landscape and the Shaping of English National Identity* (Cambridge: Cambridge University Press, 2018), 152.
49 See for example Wainwright, *A Pictorial Guide, Book One*, Helvellyn 21, Hart Side 6; *Book Four*, Lingmell 7; *Book Five*, Great Calva 3, Skiddaw 3; *Book Six*, Sale Fell 3–4.
50 Derek Cockell, 'Literary References in AW's Guides', *Footsteps*, 77 (2022), 8–10; 'Literary References in AW's Guides (Part 2)' *Footsteps*, 78, (2022), 12–3. Wainwright referenced locations in Wordsworth's poems: *The Excursion* (Bleatarn House), *Song at the Feast of Brougham Castle* (Bowscale Tarn) and *Yew Trees* (the Borrowdale Yews).
51 Saeko Yoshikawa, *William Wordsworth and the Invention of Tourism, 1820–1900* (Farnham: Ashgate, 2014); *William Wordsworth and Modern Travel: Railways, Motorcars and the Lake District, 1830–1940* (Oxford: Liverpool University Press, 2020); Jean Turnbull, *The Impact of Motor Transport on Westmorland c.1900–39* (Carlisle: Cumberland and Westmorland Antiquarian and Archaeological Society, 2021).

especially about Everest.[52] Mike Parsons and Mary Rose have drawn attention to the successful ascent of Everest in 1953 with no fatalities as a catalyst which made mountaineering seem safe and mountaineers influential to the wider public.[53] In *Book One* Wainwright suggested that 'Everest enthusiasts will liken the two pronounced rises' of High and Low Pike to the 'first and second steps' of the mountain's north-east ridge.[54] There were, however, limits to exploration for Wainwright, as for many of the guidebook writers who preceded him. Wainwright described the best places for a fell-walker to find excitement without danger – and without rock climbing.[55] The origins of the growing distinction between fell-walking and other forms of upland pedestrianism are found in the later nineteenth century, and are informed and reflected by the guidebooks of Prior, Jenkinson and Baddeley. Increased danger had been associated with climbing after the Matterhorn tragedy of 1865 which left dead four climbers and guides in Edward Whymper's party. This association ultimately led to the codification of climbing as a sport in the later years of the nineteenth century. Improved mapping may have allowed Prior, Jenkinson and Baddeley to suggest some intrepid routes for walkers, but they cautioned their readership against rock climbing. Wainwright echoed their guidance, strongly advising against climbing Pillar Rock: 'Don't even try to get a foothold on it. The climbing guides mention easy routes [...] but these are NOT easy for a walker who is not a climber.'[56] The personal sufferings Wainwright experienced during the ascent of Skew Gill on Great End attached to a rope in all likelihood influenced his advice.[57] The author and climber A. Harry Griffin recollected a separate incident when Wainwright regarded

52 Davies, *Biography*, 89.
53 Mike C. Parsons and Mary B. Rose, *Invisible on Everest* (Philadelphia: Northern Liberties Press, 2002), 216–7; 220.
54 Wainwright, *A Pictorial Guide, Book One*, High Pike 1.
55 Wainwright, *A Pictorial Guide, Book Seven*, Some Personal Notes in Conclusion.
56 Wainwright, *A Pictorial Guide, Book Seven*, Pillar 5.
57 Wainwright, *A Pictorial Guide, Book Four*, Great End 12.

a rope which Griffin had produced from his rucksack as if it were a poisonous snake.[58]

Wainwright's signature appears after the *Pictorial Guides'* extended title, pointing to their personal nature, which helped Wainwright to become a companion in the pocket of the reader-walker intent on exploring the uplands of the English Lake District. Wainwright's status as companion represented a culmination of the process seen in earlier works where guidebooks gradually superseded hiring a local guide, as well as an attempt to direct the minds as well as the feet of readers. Wainwright expressed an ethos of fell-walking which shifted the reader-walker away from more standard guidebook fare by drawing the like-minded into his way of appreciating and exploring the landscape. A good walker moved firmly with a smooth and rhythmical stride, tidily without displacing loose stones, did not cut corners by hurrying, and preferred to be solitary and quiet in order better to explore and respect his surroundings.[59] The good walker was also an explorer, like himself, of 'unfrequented corners, zigzagging where there is no need to zigzag, sometimes returning to the same summit two or three times during the course of a day' in contrast to the hurrier, the cutter of corners.[60] Wainwright stated that 'A good walker's special joy is zigzags which he follows faithfully. A bad walker's special joy is in shortcutting and destroying zigzags.'[61] Similar distaste for visitors hurrying through the district can be traced to Joseph Budworth in 1792, and for the wrong sort of visitors to Wordsworth's opposition to the extension of the railway to Windermere in 1844.[62] Wordsworth held that the 'common minds' of 'uneducated persons' brought into the district by train would not be

<hr>

58 David Johnson, *Encounters with Wainwright* (Milnthorpe: The Wainwright Society, 2016), 77.
59 Wainwright, *Memoirs of a Fellwanderer*, 113, 108–9.
60 Davies, *Letters*, 96.
61 Wainwright, *A Pictorial Guide, Book Seven*, Great Gable 16.
62 Joseph Budworth, *A Fortnight's Ramble to the Lakes in Westmoreland, Lancashire, and Cumberland. By A Rambler* (London: Hookham and Carpenter, 1792) xxvi–vii.

capable of comprehending its natural beauty except as 'an object of disgust'.[63] Wainwright's commendation of the Victorian and pre-Victorian eras in terms which could tend to hyperbole represented an attempt to guide his readers in appreciating the district correctly. He equated the merits of unfashionable Victorian viewpoints such as Glenridding Dodd, Ladies Table on Sale Fell and Lanthwaite Hill with the rightness of Victorian values, reinforced in this description of Bowscale Tarn: 'Yet the Victorian travellers were right, their sense of values was always sound. The setting is wild and romantic and very impressive.'[64] His praise for the unspoiled and 'sturdily independent' inhabitants of the Northern Fells echoed the 'perfect Republic of Shepherds and Agriculturists' extolled by Wordsworth.[65]

Wainwright's extensive explorations on foot led him to develop a distinct upland aesthetic based on the practice of fell-walking. This aesthetic, combined with his stance on issues concerning the use of the resources of the LDNP and his appreciation of the human and historical past of the district, aligned him broadly with the environmental approaches to the landscape of the area that Readman has traced in Wordsworth's *Guide*.[66] Clare Palmer and Emily Brady have defined Wainwright's upland aesthetic as being a mixture of the Picturesque, favouring height, roughness and rockiness over the smooth and regular, and the Romantic, represented by the longing to be overawed by nature's power in the presence of the sublime.[67] A notable characteristic was a dislike for upland bogs, also shared by Baddeley. Palmer and Brady suggest this dislike arose because such features

63 William Wordsworth, *Guide to the Lakes* ed. by Ernest de Selincourt, reprint of 5th edn (Oxford: Oxford University Press, 1970), 151

64 Wainwright, *A Pictorial Guide, Book One*, Glenridding Dodd 1; *Book Six*, Sale Fell 5; Grasmoor 3; *Book Five*, Bowscale Fell 7.

65 Wordsworth, *Guide to the Lakes*, 67; Wainwright, *A Pictorial Guide, Book Five*, Some Personal Notes in Conclusion.

66 Readman, *Storied Ground*, 152.

67 Clare Palmer & Emily Brady, 'Landscape and Value in the Work of Alfred Wainwright (1907–91)' *Landscape Research* Vol 32/4 (2007), 397–421; 400–8, accessed 25 March 2019.

made the landscape harder to traverse on foot.[68] Wainwright deplored the detrimental effects of man on the environment of the English Lake District caused by the motor car, water extraction, nuclear power and insensitive afforestation. His views put him in tune with amenity groups, such as the Council for the Preservation of Rural England (CPRE) and the Friends of the Lake District (FLD), and with the founder of the latter Rev. H. H. Symonds, author of *Walking in the Lake District* (1933).[69]

Fell-walking for Wainwright was an embodied practice which not only influenced his landscape values but had a moral dimension beyond the discernment of a certain taste and type of visitor. His fell-walking ethos distinguished between those purposively exploring the fells, using a rich vocabulary which included strong and/or adventurous walkers, true fell walkers, wanderers, botanists, hikers, pilgrims, explorers, scramblers, climbers and travellers; and others whom he described variously as peak baggers, tourists, vandals, wreckers, litterers, and motorists. He presented the first group as more deserving than tourists and their proxies in a way which mirrors a distinction Buzard has highlighted in Wordsworth's poem *The Brothers*. Wordsworth contrasted the purposeful traveller of 'twelve stout miles' with the tourist flitting about like a butterfly or sitting and scribbling.[70] For Wainwright the purpose of walking and how the walker moved through the landscape were key both in fell-walking and in life.

---

68 Palmer & Brady, 'Landscape and Value in the Work of Alfred Wainwright', 399; 405. Baddeley, *Thorough Guide*, 198 describes Blea Tarn near Ullscarf as 'a dismal sheet of water in the midst of a peat-bog'. Wainwright, *A Pictorial Guide, Book Three*, Ullscarf 13 refers to the route between Ullscarf and Armboth Fell passing Blea Tarn as 'one of the wettest walks in Lakeland'.

69 Geoffrey Berry and Geoffrey Beard, *The Lake District, A Century of Conservation* (Edinburgh: John Bartholomew & Son, 1990), 14. The FLD was founded in 1934 by Symonds partially in response to the Forestry Commission's acquisition of land in Eskdale in 1933. By September of that year the Forestry Commission had planted nearly one and a quarter million larch and over five million spruce in its estates in Ennerdale and Thornthwaite near Keswick. The FLD and CPRE favoured the creation and preservation of native broadleaved woodland over the Forestry Commission's planting of larch and spruce.

70 Buzard, *Beaten Track*, 20.

He stated somewhat uncompromisingly that 'Mountain climbing is an epitome of life, and good practice for it. You start at the bottom, the weaklings and the irresolute drop out on the way up, the determined reach the top. Life is like that.'[71] Wainwright equated the smooth passage of a good walker with life's virtues and the clumsiness causing the trail of debris left by a bad walker with life's vices.[72] Symonds too had previously made walking into a subject for philosophy. He described mountain walking in its most elevated form as a practice which combined skills, interests and knowledge spanning architecture, local customs and the sciences of geology and botany.[73]

Wainwright's fell-walking ethos reflected issues of inclusivity and exclusivity which can be traced in preceding guidebooks to the district. Similar distinctions to those he made in his readership are seen in Prior's and Symonds's disparagement of visitor practices which they defined as vulgar, as well as Wordsworth's earlier opposition to the railways.[74] Prior addressed his guidebook to an exclusive and experienced audience of Alpinists or those aspiring to be. For Symonds, the value of the 'intelligence' and 'art' of walking culminated in the transcendental and masculinist experience of being part of a 'brotherhood' at one with a universe created by God.[75] Wainwright tended to emphasize the masculine over the feminine and to adopt a derogatory tone about certain women, a part of contemporary culture and one which may be read as tongue-in-cheek. His suggestion that some wives might meet a deserved death on Wetherlam because their husbands had tired of them hints at his own unhappy first marriage. Women wearing skirts descending Glaramara were at least given the power to reject male gallantry.[76] His observation 'that fancy handbags and painted toenails are as likely to be seen as rucksacks and boots' on

71 Wainwright, *Memoirs of a Fellwanderer*, 194.
72 Wainwright, *Memoirs of a Fellwanderer*, 113.
73 H. H. Symonds, *Walking in the Lake District*, pocket edition (London: Maclehose, 1935), 252–3.
74 Prior, *Ascents and Passes*, 19; Symonds, *Walking in the Lake District*, 39; 221.
75 Symonds, *Walking in the Lake District*, 49, 58–9.
76 Wainwright, *A Pictorial Guide, Book Four*, Glaramara 7.

Coniston Old Man can be read as a comment on the character of the fell itself. The illustration of the crowded summit is, however, significant, Wainwright praising the solitary male fell-walker gazing towards the hills.[77] Evidence from Wainwright's life adds more nuance. Although only two of the twenty-four *Pictorial Guides'* letters reproduced in Hunter Davies's *Wainwright Letters* are from women, Wainwright enjoyed regular correspondence with others, notably the writer Molly Lefebure.[78] David Johnson's *Encounters with Wainwright* has further redressed this imbalance. Many contributors highlight Wainwright's individual acts of personal kindness and encouragement in their fell-walking endeavours.[79] By contrast, Jenkinson's *Practical Guide* expressed greater inclusiveness. He 'travelled on foot over almost every inch of the ground' for his guidebook to be 'of service to the tourist' and to make the pleasures he found in fell-walking accessible to all who were able to enjoy them.[80] He advocated public access to the fells and led the Latrigg Fell Mass Trespass of 1887, which aimed to maintain the right of pedestrian access to this popular, but modest, fell of 1,203 feet overlooking the town of Keswick.

The initial reception and consumption of the *Pictorial Guides* was positive. Enthusiastic fan letters started to arrive almost immediately, some suggesting minor corrections. Wainwright referred to having already received several hundred letters by 1958 when Book Three was published and he was working on Book Four.[81] Sales built initially through word of mouth, assisted by favourable reviews which appeared predominantly in local publications, though the *Manchester Guardian* reproduced an ascent diagram in June 1955. Bill Mitchell, the editor of *Cumbria* magazine, included a piece on each book as it was published.[82] Other aspects of the *Pictorial Guides'* reception pointed to ongoing tensions between the

---

77 Wainwright, *A Pictorial Guide, Book Four*, Wetherlam 8; Coniston Old Man 8; 13.
78 Davies, *Letters*, 97–124. Molly Lefebure wrote on the Lake District and the Lake Poets and became a friend of Wainwright, who illustrated two of her books for children.
79 Johnson, *Encounters with Wainwright*, 87–8, 113–5, 176–86.
80 Jenkinson, *Practical Guide*, iii.
81 Davies, *Letters*, 101.
82 Davies, *Biography*, 158–9, 168.

preservation and protection of the LDNP and suggested that for some readers these guidebooks promoted over-reliance. John Wyatt, the first Warden of the LDNP, claimed in 1966 that they were already out of date, had over-popularized the area and were to blame for a number of mountain accidents. Wainwright rejected this criticism, claiming instead that 'my books have often saved people from benightment and injury, and that there would be more incidents without them.'[83] Likewise, the erosion of footpaths was not caused by the growth in the popularity of fell-walking as a pastime, but by 'clumsy walkers', particularly those in groups who extended the boundaries of the trodden ways by walking abreast to converse.[84] The debate between providing information and guidance and allowing personal choice and responsibility continues today as more people take to the uplands of the region. The website of Keswick Mountain Rescue Team states that 'Occasionally paths and rights of way are badly placed on OS maps' but names Wainwright as a 'culprit' in some rescues. The popularity of his *Pictorial Guides* has led some walkers to go astray or underestimate the difficulties of the lesser heights of Barf and Catbells.[85] By contrast, Griffin suggested that, while Wainwright's guidebooks were remarkable, they were too complete and risked taking away the joy of discovery of the fells.[86]

Wainwright's television appearances in the 1980s brought the *Pictorial Guides* to the attention of a new audience. He had previously declined requests for publicity, his reply to questions posed by Mitchell on the publication of Book One from which the title of this chapter is derived remaining unsent.[87] The motivation for his television appearances was to donate funds to Animal Rescue Cumbria, to whom he also gave the royalties from

83 Davies, *Letters*, 124.
84 Wainwright, *Memoirs of a Fellwanderer*, 179–80.
85 <https://keswickmrt.org.uk/safety-hotspots/> In 2018 the team attended three separate incidents in two weeks where people had followed an erroneous path above Grange-in-Borrowdale, two of which led to falls and injury.
86 Davies, *Biography*, 158.
87 Davies, *Letters*, 93–6.

his books.[88] Wainwright's publisher, Andrew Nichol, has recorded that sales of all the *Westmorland Gazette's* Wainwright titles including the *Pictorial Guides* tripled from 28,284 in 1982 to 87,378 in 1986 when the author had become a household name.[89] After Wainwright's death in 1990, annual sales in 1992 were 25,000 copies of the seven *Guides*.[90] Average sales of 3,500 copies per volume, derived by dividing the annual total sold by seven, is nearly ten times that of one of Richards's first-generation *Fellranger* guide-books.[91] Publicity brought renewed attention to issues of footpath erosion and consideration of Wainwright's character and views. The latter were commonly framed within the caricature of the curmudgeonly Northerner to which print media has frequently defaulted to describe him but are at odds with the many personal recollections of his shyness and generosity described in Johnson's *Encounters with Wainwright*.[92]

If the *Pictorial Guides* had been more conventionally designed and published it is easy to see how they may have been received and potentially anchored in time as being the guidebooks of the newly formed LDNP. Instead, Wainwright's guidebooks remain relevant today, inspiring tourism and upland exploration. They have been received as practical manuals of instruction and works of art and literature consumed on the fells, at home or on holiday. Fell-walking has become an important leisure activity in a nationally and globally significant landscape, the LDNP having been designated an UNESCO World Heritage Site in 2017. *The Wainwright*

88 Davies, *Biography*, 267–8.
89 Davies, *Biography*, 297.
90 Davies, *Biography*, 333.
91 Sales varied between books, Book Four *The Southern Fells* in the late 1980s having been the best-selling volume for a number of years. Joe Williams of Cicerone has kindly provided the following context for these sales figures. Assuming an average figure of 3,500 copies per volume sold (by dividing the annual figure by seven), each guidebook would be within Cicerone's top five bestsellers in 2019, representing very strong sales. Mark Richards's Fellranger series of eight guides (published by Cicerone) covering the entire region were completed in 2013 and have since been redesigned, maintaining an ascent narrative and structure of one chapter per fell influenced by Wainwright.
92 Ged Moran, 'Alfred Wainwright Fogy of the Fells', *The Guardian* (27 October 1990); Martin Kettle, 'Nasty Side of a Great Feller', *The Guardian* (20 September 1994).

Society promotes Wainwright's walking ethos, in maintaining lists of those who complete the 214 fells and other Wainwright challenges, and in republishing some out-of-print titles, while Wainwright first editions, memorabilia and drawings are highly collectable.[93] From February to December 2025, The Armitt Library and Museum in Ambleside co-curated, with Wainwright archivist Chris Butterfield, a special exhibition to mark the seventieth anniversary of the publication of Book One. The *Pictorial Guides* have also inspired other literary projects, notably Andy Beck's *The Wainwrights in Colour,* and personal accounts of and guidebooks on completing the Wainwright fells.[94] Records have been established for the quickest continuous and non-continuous rounds of the 214 fells. The determination and resolution required to design and complete these endurance challenges speak to Wainwright's connection of good fell-walking with the qualities needed to 'reach the top' in life, though not to his preference for leisurely exploration.

In his personal notebook Wainwright knitted together his explorations of the uplands of the English Lake District with his knowledge of its literature and aspects of the content of previous guidebooks which informed and reflected changing ways in which tourists experienced the fells on foot. A design rooted in Wainwright's interests and skills, combined with his awareness of the print history of the district, help to make the *Pictorial Guides* a successful example of printing for tourists and lends them an appeal which extends beyond this particular form of consumption. His fell-walking ethos realizes, more fully than previous guidebooks, the persona of the author as a guide in the reader-walker's pocket. The companionship the *Pictorial Guides* induce pays homage to the roots of fell-walking and allows the reader-walker to be accompanied both on the fells and elsewhere by a guide in the form of a guidebook.

93 A pen and ink sketch of Striding Edge was auctioned for a record price of £10,200 in June 2020. <https://www.thewestmorlandgazette.co.uk/news/18560499.alfred-wainwrights-sketch-sells-record-price-online-auction/>

94 Andy Beck, *The Wainwrights in Colour* (Bowes, County Durham: Double Z, 2017); Andy Grigg, *Nowhere Fast, Walking the Wainwrights* (Ammanford: Sigma, 2014); Graham Uney, *Walking the Wainwrights* (Caernarfon: Pesda Press, 2021).

Notes on Contributors

CATHERINE ARMSTRONG, PhD, is Professor of Modern History at Loughborough University. She has co-edited many volumes on book history and is a former editor of the journal *Publishing History*. She also publishes on the history of slavery and race in the United States, on transgender history and is Director of People and Culture for the School of Social Sciences and Humanities at Loughborough University.

SARAH BURKE CAHALAN is a writer, researcher, and former librarian based in Dayton, Ohio. From 2016 to spring 2023, she served as Director of the Marian Library at the University of Dayton.

ALEXANDER DA COSTA, PhD, is Associate Professor at the University of Cambridge, where they research incunabula and early sixteenth-century printing. Their second book, *Marketing English Books* (OUP 2020) includes a chapter on 'Wide-ranging appetites: pilgrimage guides, advertisements and souvenirs' which builds on an earlier article on marketing local, English shrines by print.

ANTHONY HAMBER, PhD, is an independent photographic historian specializing in the period 1839–80. He has authored several books on a variety of aspects of nineteenth-century photographic practice.

KAYLA HARRIS is the Director of the Marian Library at the University of Dayton, one of the largest collections of materials on the Blessed Virgin Mary in the world. She has a masters degree in library and information science, and has recently written on devotion at Marian shrines through the internet.

ELAINE JACKSON, PhD, is an independent researcher, particularly interested in book history, bibliography and women's writing. She has contributed to the *Virginia Woolf Bulletin, Diegesis: Journal of the Association for Research in Popular Fictions*, the *Encyclopaedia of British Women's Writing 1900–1950* (2005), *Book Trade Connections from the Seventeenth to the Twentieth Centuries* (2008) and is co-editor of *Transient Print: Essays on the History of Printed Ephemera* (2023).

STEVEN F. JOSEPH is a Brussels-based historian of photography. Author of numerous articles and books on nineteenth-century photography in Belgium, he also researches the application of early photography to book illustration.

IAN MAXTED was from 1977 to 2005 the local studies librarian for Devon, based in the Westcountry Studies Library in Exeter. He was responsible for the 'Etched on Devon's Memory' project to digitize as many as possible of the 3,500 topographical prints recorded by John Somers Cocks in 2004/5. He is the compiler of the Exeter Working Papers in Book History and the Devon Bibliography websites and has recently published *Dictionnaire des imprimeurs, libraries et gens du livre en Basse-Normandie* (Geneva: Droz, 2020) and *The story of the book in Exeter and Devon* (Exeter Working Papers in Book History, 2021).

SUSAN MAY, PhD, is an independent art and design historian, who takes an interdisciplinary approach, embracing visual culture, socio-economic history, politics, theology, philosophy and the history of ideas. Her interest in manuscripts and incunabula was stimulated by her PhD topic, the iconographical programme of the Piccolomini Library, Siena (2006); the research included studying the library's extant bibliographic contents. She has published detailed analyses of paintings, sculpture, architecture, civic ritual, cardinalate patronage and the printed word and image.

KAREN MCAULAY is a Fellow of the Higher Education Academy, and is a postdoctoral research fellow combining musicology with cultural, library and book history, at the Royal Conservatoire of Scotland. Karen worked on the AHRC-funded HMS.scot and was Principal Investigator for the

'Claimed From Stationers' Hall' network. Her book, *Our Ancient National Airs: Scottish Song Collecting from the Enlightenment to the Romantic Era* (2013), was followed by chapters in *Understanding Scotland Musically* (2018), *Music by Subscription* (2022) and *A Social History of Amateur Music Making and Scottish National Identity: Scotland's Printed Music, 1880–1951* (2025).

ANTHONY QUINN FRSA is a journalist, author and lecturer. He founded Magforum.com, the magazine history website, in 2001. His career includes being group editor at BBC Magazines/Redwood Publishing; head of publishing at West Herts College in Watford; and chief sub-editor at the *Financial Times*. Since 2015 he has been a freelance for the *Sunday Times* and a blockchain engineering company. He has written several books, and most recently contributed to *The Cambridge History of the Book in Britain*, vol 7 (CUP, 2021).

ELIF VOZAR, PhD, is a management studies researcher with experience in qualitative and quantitative data analysis. Currently Academic Program Specialist II at the University of Florida, she previously taught at the University of Exeter and the Berlin School of Business and Innovation. Her work has appeared in the *Journal of Humanities and Applied Social Sciences* and *CABI Tourism Cases.*.

THOMAS M. VOZAR, PhD, is Assistant Professor of Humanities at the Hamilton School for Classical and Civic Education, University of Florida. His books include *Milton, Longinus, and the Sublime in the Seventeenth Century* (Oxford University Press, 2023), which won the Milton Society of America's James Holly Hanford Book Award, and *Isaac Barrow's On the Turkish Religion: A Latin Poem on Islam from Ottoman Istanbul* (Bloomsbury, 2026).

LIZ WOODHAM is a PhD researcher at Lancaster University studying how English Lake District pocket guidebooks and the leisure practice of fell-walking in the area have shaped each other from the mid-nineteenth to the mid-twentieth centuries. Her thesis addresses an element of the identity of the English Lake District by telling the story of the way in which pocket guidebooks increase access to the landscape by outsiders.

SHIJIA YU, PhD, is Associate Lecturer at the Open University. Drawing on media archaeology and material culture studies, Shijia's research interest is in how the paper peepshow can be used as a tool to expand our understanding of nineteenth-century visual entertainments and print novelties, as well as the use of embodied knowledge in visual culture studies. She has recently written on the paper peepshow and the Thames Tunnel (built. 1843), published in The Home, Nations and Empire, and Ephemeral Exhibition Spaces: 1750-1918 (Amsterdam U Press, 2021).

Printing History and Culture

Series Editors

Caroline Archer-Parré, Malcolm Dick and Hazel Wilkinson

This series unites the allied fields of global, national and local printing history and print culture, and is therefore concerned not only with the design, production and distribution of printed material but also its consumption, reception and impact. It includes the histories of the machinery and equipment, of the industry and its personnel, of the printing processes, the design of its artefacts (books, newspapers, journals, fine prints, and ephemera) and with the related arts and crafts, including calligraphy, type-founding, typography, papermaking, bookbinding, illustration, and publishing. It also covers the cultural context and environment in which print was produced and consumed.

This series is issued by the Centre for Printing History and Culture, a joint initiative between Birmingham City University and the University of Birmingham that seeks to encourage research into all aspects and periods of printing history and culture, as well as education and training into the art and practice of printing.

Published Volumes

Vol. 1   Ian Cawood and Lisa Peters (eds): *Print, Politics and the Provincial Press in Modern Britain.*
2019. 260 pages. ISBN 978-1-78874-430-0

Vol. 2   Artemis Alexiou and Rose Roberto (eds): *Women in Print 1: Design and Identities.*
2022. 320 pages. ISBN 978-1-78997-978-7

Vol. 3   Caroline Archer-Parré, Christine Moog and John Hinks (eds): *Women in Print 2: Production, Distribution and Consumption.*
2022. 296 pages. ISBN 978-1-78997-977-0

Vol. 4   Caroline Archer-Parré and James Mussell (eds): *Letterpress Printing: Past, Present, Future.*
2023. 300 pages. ISBN 978-1-80079-421-4

Vol. 5   Lisa Peters and Elaine Jackson (eds): *Transient Print: Essays on the History of Printed Ephemera.*
2023. 276 pages. ISBN 978-1-78997-900-8

Vol. 6   Sahar Afshar, Wei Jin Darryl Lim and Vaibhav Singh (eds): *Script, Print and Letterforms in Global Contexts: The Visual and the Material.*
2025. 226 pages. ISBN 978-1-78997-503-1

Vol. 7   Catherine Armstrong and Elaine Jackson (eds): *Print and Tourism: Travel-Related Publications from the Sixteenth to the Twentieth Century.*
2026. 276 pages. ISBN 978-1-80374-244-1